Discrete Structures
and Their
Interactions

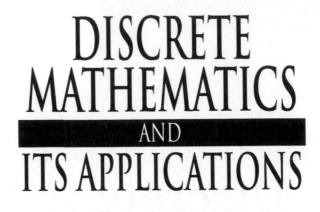

DISCRETE MATHEMATICS AND ITS APPLICATIONS

Series Editor

Kenneth H. Rosen, Ph.D.

Titles (continued)

Jacob E. Goodman and Joseph O'Rourke, Handbook of Discrete and Computational Geometry, Second Edition

Jonathan L. Gross, Combinatorial Methods with Computer Applications

Jonathan L. Gross and Jay Yellen, Graph Theory and Its Applications, Second Edition

Jonathan L. Gross and Jay Yellen, Handbook of Graph Theory

David S. Gunderson, Handbook of Mathematical Induction: Theory and Applications

Richard Hammack, Wilfried Imrich, and Sandi Klavžar, Handbook of Product Graphs, Second Edition

Darrel R. Hankerson, Greg A. Harris, and Peter D. Johnson, Introduction to Information Theory and Data Compression, Second Edition

Darel W. Hardy, Fred Richman, and Carol L. Walker, Applied Algebra: Codes, Ciphers, and Discrete Algorithms, Second Edition

Daryl D. Harms, Miroslav Kraetzl, Charles J. Colbourn, and John S. Devitt, Network Reliability: Experiments with a Symbolic Algebra Environment

Silvia Heubach and Toufik Mansour, Combinatorics of Compositions and Words

Leslie Hogben, Handbook of Linear Algebra

Derek F. Holt with Bettina Eick and Eamonn A. O'Brien, Handbook of Computational Group Theory

David M. Jackson and Terry I. Visentin, An Atlas of Smaller Maps in Orientable and Nonorientable Surfaces

Richard E. Klima, Neil P. Sigmon, and Ernest L. Stitzinger, Applications of Abstract Algebra with Maple™ and MATLAB®, Second Edition

Richard E. Klima and Neil P. Sigmon, Cryptology: Classical and Modern with Maplets

Patrick Knupp and Kambiz Salari, Verification of Computer Codes in Computational Science and Engineering

William Kocay and Donald L. Kreher, Graphs, Algorithms, and Optimization

Donald L. Kreher and Douglas R. Stinson, Combinatorial Algorithms: Generation Enumeration and Search

Hang T. Lau, A Java Library of Graph Algorithms and Optimization

C. C. Lindner and C. A. Rodger, Design Theory, Second Edition

San Ling, Huaxiong Wang, and Chaoping Xing, Algebraic Curves in Cryptography

Nicholas A. Loehr, Bijective Combinatorics

Toufik Mansour, Combinatorics of Set Partitions

Alasdair McAndrew, Introduction to Cryptography with Open-Source Software

Elliott Mendelson, Introduction to Mathematical Logic, Fifth Edition

Alfred J. Menezes, Paul C. van Oorschot, and Scott A. Vanstone, Handbook of Applied Cryptography

Titles (continued)

Stig F. Mjølsnes, A Multidisciplinary Introduction to Information Security

Jason J. Molitierno, Applications of Combinatorial Matrix Theory to Laplacian Matrices of Graphs

Richard A. Mollin, Advanced Number Theory with Applications

Richard A. Mollin, Algebraic Number Theory, Second Edition

Richard A. Mollin, Codes: The Guide to Secrecy from Ancient to Modern Times

Richard A. Mollin, Fundamental Number Theory with Applications, Second Edition

Richard A. Mollin, An Introduction to Cryptography, Second Edition

Richard A. Mollin, Quadratics

Richard A. Mollin, RSA and Public-Key Cryptography

Carlos J. Moreno and Samuel S. Wagstaff, Jr., Sums of Squares of Integers

Gary L. Mullen and Daniel Panario, Handbook of Finite Fields

Goutam Paul and Subhamoy Maitra, RC4 Stream Cipher and Its Variants

Dingyi Pei, Authentication Codes and Combinatorial Designs

Kenneth H. Rosen, Handbook of Discrete and Combinatorial Mathematics

Douglas R. Shier and K.T. Wallenius, Applied Mathematical Modeling: A Multidisciplinary Approach

Alexander Stanoyevitch, Introduction to Cryptography with Mathematical Foundations and Computer Implementations

Jörn Steuding, Diophantine Analysis

Douglas R. Stinson, Cryptography: Theory and Practice, Third Edition

Roberto Togneri and Christopher J. deSilva, Fundamentals of Information Theory and Coding Design

W. D. Wallis, Introduction to Combinatorial Designs, Second Edition

W. D. Wallis and J. C. George, Introduction to Combinatorics

Jiacun Wang, Handbook of Finite State Based Models and Applications

Lawrence C. Washington, Elliptic Curves: Number Theory and Cryptography, Second Edition

DISCRETE MATHEMATICS AND ITS APPLICATIONS

Series Editor KENNETH H. ROSEN

Discrete Structures *and Their* Interactions

Jason I. Brown

CRC Press
Taylor & Francis Group
Boca Raton London New York

CRC Press is an imprint of the
Taylor & Francis Group, an **informa** business

A CHAPMAN & HALL BOOK

CRC Press
Taylor & Francis Group
6000 Broken Sound Parkway NW, Suite 300
Boca Raton, FL 33487-2742

First issued in paperback 2019

© 2013 by Taylor & Francis Group, LLC
CRC Press is an imprint of Taylor & Francis Group, an Informa business

No claim to original U.S. Government works

ISBN-13: 978-1-4665-7941-5 (hbk)
ISBN-13: 978-0-367-37985-8 (pbk)

Visit the Taylor & Francis Web site at
http://www.taylorandfrancis.com

and the CRC Press Web site at
http://www.crcpress.com

Contents

List of Figures

Symbol Description

X^n	Cartesian product of the set X with itself n times	$G - e$	deletion of edge e in graph G
$\mathcal{P}(X)$	power set of X	$G \bullet e$	contraction of edge e in graph G
$\mathbb{N}$	set of natural numbers (starting at 1)	$\tau(G)$	vertex cover number of graph G
$\mathbb{Z}$	set of integers	$\overline{G}$	complement of graph G
$\mathbb{Z}_{\geq 0}$	set of nonnegative integers	$cl(G)$	clique partition number of graph G
$\mathbb{Q}$	set of rational numbers	R^*	transitive closure of relation R
$\mathbb{R}$	set of real numbers	$x||y$	x is incomparable to y
$\mathbb{C}$	set of complex numbers	$\bigvee S$	join of vertices in S
P	class of polynomial problems	$\bigwedge S$	meet of vertices in S
NP	class of nondeterministically polynomial problems	$D(v)$	ideal generated by v
		$U(v)$	filter generated by v
NPc	class of nondeterministically polynomial complete problems	Indisc_n	indiscrete topology on $[n]$
		Disc_n	discrete topology on $[n]$
#-P	class of nondeterministically polynomial counting problems	$\mathrm{Cov}(P)$	covering graph of poset P
		Top_V	collection of topologies on set S
$\deg_S(v)$	degree of vertex v in discrete structure S	Pre_V	collection of preorders on set V
$\delta(S)$	minimum degree of vertex v in discrete structure S	$\bigwedge^k V$	k^{th} exterior (or alternating) power of V
$\Delta(S)$	maximum degree of vertex v in discrete structure S	K_n^l	complete l–uniform hypergraph of order n
E_n	empty graph of order n	Δ	a complex
K_n	complete graph of order n	$\mathrm{del}_\Delta(v)$	deletion complex
P_n	path of length n	$\mathrm{link}_\Delta(v)$	link complex
C_n	cycle of length n	$G(n,k)$	Kneser graph
$K_{n,m}$	complete bipartite graph with cells of cardinality n and m.	$k[\Delta]$	Stanley-Reisner ring of complex Δ over field k
$\chi(G)$	chromatic number of graph G	$\pi(G,x)$	chromatic polynomial of graph G
$\beta(G)$	independence number of graph G		
$\omega(G)$	clique number of graph G		

Preface

I'd love to turn you on ...

(A Day in the Life, Lennon & McCartney)

Imagine walking along a beach, looking for seashells. Each seashell you unearth is precious and exquisite, with its own unique beauty. Discovering the connections between different areas of mathematics is much the same. There are treasures to be found in applying one field to another in unexpected ways.

This text is intended for graduate and upper level undergraduate students in mathematics who have taken an initial course in discrete mathematics or graph theory. This text introduces a number of discrete structures (such as hypergraphs, finite topologies, preorders, simplicial complexes and order ideals of monomials) that most graduate students in combinatorics (and indeed, some researchers in the field) seldom experience. These discrete structures have important applications to many areas, both inside and outside combinatorics, and unless you aware of the structures, you will miss valuable connections that could be forged within your research.

Discrete structures can, from the right viewpoint, interact with one another, affording one the ability to use ideas and techniques that are "natural" in the second area to the first, often with striking results. On the other hand, we illustrate that discrete structures can be used sometimes even if one is interested in another seemingly unrelated area of mathematics or computer science, merely to represent the salient features and discover the underlying combinatorial principles. Finally, other areas of mathematics, such as linear and multilinear algebra, commutative algebra, topology, probability theory and logic, to name but a few, can have startling applications to problems exclusively within combinatorics, and this text highlights a few of such jewels.

This book does not attempt to delve deeply into any one area of discrete structures or to give the most recent or most involved applications in the research literature; the emphasis is on the connections between different structures and fields, and I hope that the examples chosen are accessible to most. This book is a leisurely stroll along the beach, looking for interesting approaches, ones that you might not find otherwise. I hope that after reading through this book, you will find some of your own.

Acknowledgments

I'd like to acknowledge Angela Siegel for her comments and corrections, and Sunil Nair, Rachel Holt and Prudence Board at CRC for their support. And as always, I thank my wife, Sondra, and two sons, Shael and Zane, all of whom I love *discretely* (and continuously).

About the Author

Jason I. Brown received his B.Sc. in mathematics in 1984 from the University of Calgary, where he was caught up in the vortex of discrete mathematics circling there at the time (including visits from the legendary Paul Erdös). He then went on to the University of Toronto where he received his M.Sc. (1985) and Ph.D. (1987). He is presently a professor of Mathematics at Dalhousie University on the beautiful East coast of Canada. Jason has written over 70 refereed publications. His research interests are broad, including graphs, hypergraphs, partial order, finite topologies and simplicial complexes, always with an eye on the applications of various other fields of mathematics to the discrete problem at hand.

On the side, he has interests in the connections between mathematics and music as well. His 2004 mathematical research that uncovered how the Beatles played the opening chord of *A Hard Day's Night* garnered widespread media attention, including *NPR* and *BBC* radio, *Guitar Player Magazine*, and the website of the *Wall Street Journal*. Jason is an avid guitar player and songwriter, and has recorded his first CD, *Songs in the Key of Pi*. You can find out more at *http:www.mathstat.dal.ca/brown* and *http://www.jasonibrown.com*.

Chapter 1

Introduction

There are a variety of common discrete structures – graphs, directed graphs, partial orders, lattices, hypergraphs, matroids – and some lesser known ones, at least to combinatorialists – preorders, finite topologies, simplicial complexes and multicomplexes. We are going to bring these together within a general framework, and show that while each of the discrete structures has its own merits, uses and advantages, it is the modeling of one structure by another, or by a seemingly unrelated structure elsewhere within mathematics, that often provides startlingly new insights. The alternate viewpoints suggest new modes of attack on old problems and less traveled roads to explore.

In terms of background, we assume that you are acquainted with the basics of undergraduate mathematics – calculus, linear algebra, rings and fields, set theory, first-order logic and probability theory (for more details, see the appendices). Any other topics, such as commutative and multilinear algebra and topology, will be introduced on a need–to–know basis. There are a few bits of notation that may be non-standard that we will list now.

1.1   Sets

In terms of set theory notation, we have one addition: we define a **multiset** as a set with repetitions allowed, such as $\{a, a, b, c, c, c, e\}$. One multiset is contained in another if for each element x of the first multiset, x appears at least as often in the second multiset. The usual notions of union and intersection extend in the natural way from sets to multisets, that is, they incorporate repetitions.

As usual, the sets of natural, integer, rational, real complex numbers will be denoted by $\mathbb{N}$, $\mathbb{Z}$, $\mathbb{Q}$, $\mathbb{R}$ and $\mathbb{C}$, respectively (we exclude 0 from $\mathbb{N}$). By $\mathbb{Z}_{\geq 0}$ we mean the set of nonnegative integers. Finally, as the set $\{1, 2, \ldots, n\}$ occurs frequently, we abbreviate it as $[n]$. By $\mathcal{P}(X)$ we mean the **power set** of X, that is, the set of subsets of X. For a function f on a set X and a subset S of X, $f(S)$ denotes the set $\{f(s) : s \in S\}$.

1.2 Sequences

We say that a sequence $\langle b_0, \ldots, b_d \rangle$ of real numbers is **unimodal** if there is a $k \in \{0, \ldots, d\}$ such that

$$b_0 \leq b_1 \leq \cdots \leq b_{k-1} \leq b_k \geq b_{k+1} \geq \ldots \geq b_d$$

and **log concave** if

$$b_i^2 \geq b_{i-1} b_{i+1} \quad \text{for } 0 < i < d$$

(the sequence is **strictly log concave** if strict inequality holds for all such i). It is not hard to see that for positive or alternating sequences (with no zero terms), log concavity implies unimodality in absolute value (that is, unimodality of the absolute values of the terms in the sequence), and strict log concavity implies furthermore that the sequence of absolute values of the terms has a single or double 'peak'. Various combinatorial sequences are known to be unimodal or log concave, such as the sequence of binomial coefficients $\langle \binom{n}{0}, \binom{n}{1}, \ldots, \binom{n}{n} \rangle$, the Stirling numbers of the first kind (c.f. [32]) and the coefficients of the Gaussian polynomials. In fact, in an article [106] describing O'Hara's constructive proof of the latter, Zeilberger states that "Combinatorialists love to prove that counting sequences are unimodal." Other sequences have been conjectured or proven to be unimodal or log concave, such as the absolute value of the coefficients in the chromatic polynomial [53, 72] and various sequences related to matroids (c.f. [103]).

1.3 Asymptotics

We shall need the "big O" and "little o" notation. If f and g are functions from the natural numbers $\mathbb{N}$ to $[0, \infty)$, we write

- $f = O(g)$ if for some positive constant C, $f(n) \leq Cg(n)$ for all sufficiently large $n \in \mathbb{N}$,

- $f = o(g)$ if $\lim_{n \to \infty} f(n)/g(n) = 0$, and

- $f \sim g$ if $\lim_{n \to \infty} f(n)/g(n) = 1$.

For example, if $f(n) = 2n^2 - 3\sqrt{n} + 8$, $g(n) = n^2$, $h(n) = 2n^2 + n\ln n$ and $k(n) = n^3$, then $f = O(g)$, $f = o(k)$ and $f \sim h$.

1.4 Computational Complexity

We refer the reader to [47] for a complete discussion of computational complexity. By **P, NP, and NPc** we denote the classes of polynomial, nondeterministically polynomial and nondeterministically polynomial complete problems. For the uninitiated, in broadest terms, **P** consists of problems that have an algorithm that runs in time $O(p(n))$ (**polynomial time**) for some polynomial in the input size n, while **NP** consists of problems that have a short 'proof' for a 'yes' answer (**co-NP** consists of problems that have a short 'proof' for a 'no' answer). One of the outstanding problems in theoretical computer science is whether **P** = **NP**. The class of **NPc** (**NP–complete**) problems are those **NP** problems such that if any one of them belonged to **P**, then $\mathbf{P} = \mathbf{NP}$ (in some sense, one can think of NP–complete problems as the 'hardest' ones in the class $\mathbf{NP}$). The $\#\textbf{-P complete}$ problems are counting problems that are the 'hardest' ones.

Some notions and results are developed in the exercises, so please try them! As well, throughout we touch on a number of different topics in undergraduate mathematics. Sadly, undergraduates (and graduates!) often miss out on some of these topics, so in the appendices, we mention those that are used in this text. The treatment is not by any means complete, and we encourage the interested reader to follow with the suggested references (the references are chosen by and large as easy introductions to the topics – I used most of them myself in my undergraduate studies). Just as you can't be too good looking, bright or happy, as a combinatorialist you can't know too much mathematics. Even a brief acquaintance with a different field of mathematics opens up enormous possibilities in your research that would be forever closed otherwise.

Chapter 2

Discrete Structures - A Common Framework

We begin by defining what we mean by a discrete structure. For positive integer k, a **k–ary** relation R on a set X is a subset of $X^k = \{(x_1, \ldots, x_k)\} : x_i \in X$ for all $i\}$; k is called the **arity** of R. A **discrete structure** D on a set V (called the (underlying) **vertex set**) is $(V, \{R_i\}_{i \in \mathcal{I}})$, where $\mathcal{I}$ is any index set and each R_i is an n_i–ary relation on V, for some positive integer n_i (if there is just one relation, we often omit the surrounding set brackets). We often write $(x_1, \ldots, x_k) \in D$ if $(x_1, \ldots, x_k) \in R_i$ for some i. Our definition allows for infinite discrete structures (when V is infinite), but unless otherwise stated, we assume a structure is finite.

Various classes of discrete structures arise based on the types and properties of the associated relations. Rather than proceed any further, let's consider some common discrete structures and how they fit into the definition just presented (note that in every case, one can associate a set or multiset of **edges** from which the original relations can be recovered). We define a discrete structure to be **undirected** if the edges are all sets rather than ordered tuples (otherwise, we say the structure is **directed**). We postpone the 'usual' pictures (which are geometric models of the structures) until later.

- A **directed graph** (or **digraph**) D is a discrete structure (V, E), where E (the **arc relation** or the **arc set**) is a binary relation on V (we sometimes write E as $A(D)$). The directed graph is **loopless** if E is irreflexive[1]. For an arc $a = (u, v) \in E$, we say that u and v are the

---

[1] A relation R on set X is **irreflexive** if $(x, x) \notin R$ for all $x \in X$.

endpoints of a, with u the **initial point** and v the **terminal point** of arc a. E is also called the **edge set** of D.

Example: Consider $V = \{1, 2, 3, 4\}$ and $E = \{(1, 1), (1, 2), (2, 4), (4, 2),$ $(3, 4)\}$. $D = (V, E)$ is a directed graph that is not loopless (it has a loop at vertex 1). △

- A **graph** G is a directed graph (V, R), where R is a binary symmetric[2] relation. The graph G is *loopless* (or **simple**) if R is irreflexive. We define the edge set of G to be $E = \{\{v\} : (v, v) \in R\} \cup \{\{u, v\} : (u, v) \in R\}$, and often write $G = (V, E)$ instead of $G = (V, R)$. A **multigraph** allows for the edge set to be a multiset rather than just a set (this corresponds to allowing for more than one symmetric binary relation on V).

Example: Consider $V = \{1, 2, 3, 4\}$ and $R = \{(1, 1), (1, 2), (2, 1), (2, 4),$ $(4, 2), (3, 4), (4, 3)\}$. $G = (V, R)$ is a graph that is not loopless (it has a loop at vertex 1), and G has edge set $E = \{\{1\}, \{1, 2\}, \{2, 4\}, \{3, 4\}\}$. △

- An **equivalence** Q is a discrete structure (V, R), where R is an equivalence (i.e. reflexive, symmetric and transitive[3] binary) relation on V, so that it is a graph as well. Recall that any equivalence relation partitions its underlying set into **equivalence classes** (two vertices x and y are in the same equivalence class iff $(x, y) \in R$). The edge set of Q is that of Q as viewed as a graph.

Example: Consider $V = \{1, 2, 3, 4\}$ and $R = \{(1, 1), (2, 4), (4, 2), (2, 2),$ $(3, 3), (4, 4)\}$. $Q = (V, R)$ is an equivalence with equivalence classes $\{1\}, \{2, 4\}$ and $\{3\}$. △

---

[2] A relation R on set X, is **symmetric** if $(x, y) \in R$ implies $(y, x) \in R$ for all x, $y \in X$.
[3] A relation R on set X is **transitive** if (x, y), $(y, z) \in R$ imply $(x, z) \in R$ for all x, y, $z \in X$.

- A **preorder** P is a directed graph $(V, \preceq)$, where $\preceq$ is a binary reflexive and transitive relation. We often write $x \preceq y$ instead of $(x, y) \in \preceq$. The edge set of P is that of P as viewed as a directed graph.

  Example: Consider $V = \{1, 2, 3, 4\}$ and $\preceq = \{(1, 1), (1, 2), (1, 4), (2, 2), (2, 4), (3, 2), (3, 3), (3, 4), (4, 2), (4, 4)\}$. $P = (V, \preceq)$ is a preorder. $\triangle$

- A **partial order** P (or **poset**) is a preorder $(V, \preceq)$, where $\preceq$ is anti-symmetric[4]. A **linear order** is a partial order $P = (V, \preceq)$ for which $x \preceq y$ or $y \preceq x$ for all x, $y \in V$. **Lattices** are partial orders with additional structure that we shall touch on later. The edge set is that of P as viewed as a directed graph.

  Example: Consider $V = \{1, 2, 3, 4\}$ and $\preceq = \{(1, 1), (1, 2), (2, 2), (2, 4), (3, 3), (3, 4), (4, 4)\}$. $P = (V, \preceq)$ is a partial order. $\triangle$

- A **hypergraph** (or **set system**) H is a discrete structure $(V, \{R_i : i \in \mathcal{I}\})$, where (i) the R_i have distinct arities and for all i, (ii) each R_i is closed under any permutation of its components, and (iii) $(v_1, \ldots, v_{n_i}) \notin R_i$ whenever any two components are equal. The hypergraph H is **k − uniform** if there is only one relation, R, and R is k-ary. Graphs are instances of hypergraphs. One can associate with a hypergraph H the subset $E = E(H)$ of V consisting of all subsets $S = \{v_1, \ldots, v_l\}$ of V for which $(v_1, \ldots, v_l) \in R_i$ for some i (as R_i is closed under any permutation of its components, we can safely write $\{v_1, \ldots, v_l\} \in R_i$ when we mean $(v_1, \ldots, v_l) \in R_i$); E is called the edge set of H (we sometimes allow an edge set of a hypergraph to contain the empty set). A **topology** on set V is simply a hypergraph whose edge set contains $\emptyset$, V, and is closed under finite intersections and arbitrary unions. A **design** is a hypergraph with certain regularity constraints on containment (see Chapter 5).

---

[4]A relation R on set X is **antisymmetric** if (x, y), $(y, x) \in R$ imply $x = y$ for all $x, y \in X$

Example: Consider $V = \{1, 2, 3, 4\}$, $R_2 = \{(1, 3), (3, 1), (3, 4), (4, 3)\}$ and $R_3 = \{(1, 2, 3), (1, 3, 2), (2, 1, 3), (2, 3, 1), (3, 1, 2), (3, 2, 1)\}$. $H = (V, \{R_2, R_3\})$ is a hypergraph. The edge set of H is $\{\{1, 3\}, \{3, 4\}, \{1, 2, 3\}\}$. △

- A **(simplicial) complex** $\mathcal{C}$ is a hypergraph whose nonempty edge set is closed under containment, that is, if $Y \in E(\mathcal{C})$ and $X \subseteq Y, X \neq \emptyset$ then $X \in E(\mathcal{C})$. Each element of the edge set of a complex is called a **face** (in the literature, the empty set is often added to a complex, but this addition poses no problem). There are many classes of complexes, with the most well–studied one being **matroids** (but more about this later – see page 118). The edge set of $\mathcal{C}$ is that of P as viewed as a hypergraph.

Example: Consider $V = \{1, 2, 3, 4\}$, $E = \{\{1\}, \{2\}, \{3\}, \{4\}, \{1, 2\}, \{1, 3\}, \{1, 4\}, \{2, 3\}, \{2, 4\}, \{3, 4\}, \{1, 2, 3\}, \{1, 2, 4\}\}$. $\mathcal{C}$, the hypergraph with vertex set V and edge set E, is a complex. △

- A **multicomplex** $\mathcal{M}$ on set V is a discrete structure $(V, \{R_i : \{i \in \mathcal{I}\})$, where the R_i have distinct arities, each R_i is invariant under any permutation of its components, and if $(v_1, \ldots, v_k) \in R_j$ for some j, then for any $i < k$, $(v_1, \ldots, v_i) \in R_l$ for some relation R_l. It is not hard to verify that any complex is a multicomplex. (Again, in the literature, the empty set is usually added to a multicomplex, as we shall often do.) A multicomplex may be viewed as consisting of multisets. The edge set of $\mathcal{M}$ is the collection of multisets $\{x_1, \ldots, x_l\}$, where $(x_1, \ldots, x_l) \in R_i$ for some i.

Example: Consider $V = \{1, 2, 3, 4\}$, $R_1 = \{(1), (2), (3), (4)\}$, $R_2 = \{(1, 1), (1, 2), (2, 1), (3, 3)\}$ and $R_3 = \{(1, 1, 1), (1, 2, 2), (2, 1, 2), (2, 2, 1)\}$. $\mathcal{M} = (V, \{R_1, R_2, R_3\}$ is a multicomplex; which we can write with multisets as $(V, \{\{1\}, \{2\}, \{3\}, \{4\}, \{1, 1\}, \{1, 2\}, \{2, 2\}, \{3, 3\}, \{1, 1, 1\}, \{1, 2, 2\}\}$. △

Thus our definition of a discrete structure includes most (if not all) objects that we would like to consider under the setting. This text shall explore the theory of, application of and interconnection between these structures.

2.1   Isomorphism

Isomorphic structures preserve all of the structures attributes that are independent of the labels of the underlying elements (I like to think of isomorphism as structures for the "near-sighted" – from a distance, where you can't see the labels of elements, only how they interact, they look the same). More precisely, two discrete structures $S_1 = (V_1, \{R_i\}_{i \in \mathcal{I}_1})$ and $S_2 = (V_2, \{Q_i\}_{i \in \mathcal{I}_2})$ are **isomorphic** (written as $S_1 \cong S_2$) if and only if there are bijections $f : V_1 \to V_2$ and $\phi : \mathcal{I}_1 \to \mathcal{I}_2$ such that for all $i \in \mathcal{I}_1$, R_i and $Q_{\phi(i)}$ have the same arity, say n_i, and for all $(v_1, \ldots, v_{n_i}) \in V_1^{n_i}$, $(v_1, \ldots, v_{n_i}) \in R_i$ if and only if $(f(v_1), \ldots, f(v_{n_i})) \in Q_{\phi(i)}$ (the maps f and ϕ are called **isomorphisms**). If a discrete structure has at most one relation of each arity, then we can omit the map ϕ as it is completely determined.

If two discrete structures are isomorphic, they will share the same properties; for example, a discrete structure isomorphic to a complex must be a complex as well. Note that the notion of an isomorphism is an equivalence relation on the class of all discrete structures, and hence partitions the class of all discrete structures into **isomorphism classes**.

For example, consider directed graphs $D_1 = (V_1, A_1)$, $D_2 = (V_2, A_2)$ and $D_3 = (V_3, A_3)$, where $V_1 = \{1, 2, 3, 4, 5\}$, $R_1 = \{(1, 2), (2, 3), (3, 4), (4, 5), (5, 1)\}$, $V_2 = \{a, b, c, d, e\}$, $R_2 = \{(a, c), (b, d), (c, e), (d, a), (e, b)\}$, $V_3 = \{1, 2, 3, 4, 5\}$ and $R_3 = \{(1, 4), (2, 4), (3, 5), (4, 1)\}$. You can verify that the function $f : V_1 \to V_2$ defined by $f(1) = a$, $f(2) = c$, $f(3) = e$, $f(4) = b$, $f(5) = d$ is an isomorphism between D_1 and D_2, so that $D_1 \cong D_2$. D_3 is isomorphic to neither D_1 nor D_2 as it has fewer arcs.

2.2 Substructures

Associated with any mathematical structure are **substructures**, that is, structures of the same type that are contained within the original structure. Groups have subgroups, semigroups have subsemigroups, vector spaces have subspaces, and so on. There are two general ways of forming substructures of discrete structures:

- An **induced substructure** of a discrete structure $S = (V, \{R_i\}_{i \in \mathcal{I}})$ is a discrete structure $S' = (V', \{Q_i\}_{i \in \mathcal{I}})$ of the same type, where $V' \subseteq V$ and each Q_i is the restriction[5] of R_i to V'; S' is said to be the substructure of S **induced** by V'.

- A **spanning substructure** of a discrete structure $S = (V, \{R_i\}_{i \in \mathcal{I}})$ is a discrete structure $S'' = (V, \{Q_i\}_{i \in \mathcal{I}})$ of the same type, where $Q_i \subseteq R_i$ for all $i \in \mathcal{I}$.

- A **(partial) substructure** of a discrete structure $S = (V, \{R_i\}_{i \in \mathcal{I}})$ is a spanning substructure of some induced substructure of S.

A substructure S' of S is **proper** if $S' \neq S$.

Example: Consider the complex $\mathcal{C}$ on vertex set $V = \{1, 2, 3, 4\}$ with edge set $E = \{\{1\}, \{2\}, \{3\}, \{4\}, \{1,2\}, \{1,3\}, \{1,4\}, \{2,3\}, \{2,4\}, \{3,4\}, \{1,2,3\}, \{1,2,4\}\}$. $\mathcal{D}$, the complex on V with edge set $E = \{\{1\}, \{2\}, \{3\}, \{4\}, \{1,2\}, \{1,3\}, \{1,4\}, \{2,3\}, \{3,4\}, \{1,2,3\}\}$, is a subcomplex of $\mathcal{C}$. $\mathcal{C}'$, the substructure of $\mathcal{C}$ induced by $\{1, 2, 4\}$, has edge set $\{\{1\}, \{2\}, \{4\}, \{1,2\}, \{1,4\}, \{2,4\}, \{1,2,4\}\}$. Note that the hypergraph H on V with edge set $\{\{1\}, \{2\}, \{3\}, \{4\}, \{1,2\}, \{1,3\}, \{1,4\}, \{2,3\}, \{3,4\}, \{1,2,3\}, \{1,2,4\}\}$ is not a subcomplex of $\mathcal{C}$ as H is not a complex (since it contains $\{1,2,4\}$ but not $\{2,4\}$) but it is a subhypergraph of $\mathcal{C}$, viewing $\mathcal{C}$ as a hypergraph rather than a complex. $\triangle$

---

[5]The **restriction** of a k–ary relation R on set V to $U \subseteq V$ is the relation $R \cap U^k$.

2.3   Properties, Parameters and Operations

While many discrete structures have their own unique properties, parameters and operations of interest, there are a few that we can place in the general setting.

Let S be a discrete structure on vertex set V with edge set E. The **order** of S is $|V|$ while its **size** is $|E|$ (note that the size depends on how the structure is viewed). We define the **degree** of vertex v in S as

$$\deg_S(v) = |\{(v, e) : v \text{ is a component of } e \in E\}|$$

(for graphs, we count loops twice). The maximum and minimum degrees of S are denoted by $\Delta(S)$ and $\delta(S)$, respectively; these are **parameters** of the discrete structure. S is said to be **k–regular** if $\Delta(S) = \delta(S) = k$, that is, every vertex has degree k.

For vertices u and v of a discrete structure S, a u–v **walk** of length k in S is an alternating sequence of vertices and edges $u = v_0, e_1, v_1, e_2, \ldots, e_k, v_k = v$ of S such that $v_i, v_{i+1} \in e_i$ for $i = 0, \ldots, k - 1$ (if S is a directed graph, then we insist that u be the initial vertex of e_1 and v the terminal vertex of e_k). For every vertex v there is a v–v walk of length 0 from v to itself. It is easy to check that the existence of a walk ("**connectedness**") is an equivalence relation on S, and the equivalence classes under this relation are called **connected components** of S (a discrete structure is **connected** if it has only one connected component). There are no edges between vertices in different connected components.

Example: Consider the hypergraph H on $V = \{1, 2, 3, 4, 5, 6\}$ with edge set $\{\{1, 3\}, \{1, 2, 4\}, \{5, 6\}\}$. Vertex 1 has degree 2, while vertex 2 has degree 1. The maximum degree $\Delta = 2$ while the minimum degree $\delta = 1$, so the hypergraph is not regular. The sequence $4, \{1, 2, 4\}, 1, \{\{1, 3\}, 3$ is a 1–4 walk of length 2 in H. Hypergraph H has two connected components, namely $\{1, 2, 3, 4\}$ and $\{5, 6\}$. $\triangle$

There are, of course, an enormous number of operations specific to each type of discrete structure. Here are two that are common to all:

- Given discrete structures $S_1 = (V_1, \{R_i\}_{i \in \mathcal{I}_1})$ and $S_2 = (V_2, \{Q_i\}_{i \in \mathcal{I}_2})$ with $V_1 \cap V_2 = \emptyset$, the **disjoint union** of S_1 and S_2 is the discrete structure $S_1 \cup S_2 = (V_1 \cup V_2, \{R_i\}_{i \in \mathcal{I}_1} \cup \{Q_i\}_{i \in \mathcal{I}_2})$. If $V_1 \cap V_2 \neq \emptyset$, then we take disjoint isomorphic copies before we proceed.

- Given discrete structures $S_1 = (V_1, \{R_i\}_{i \in \mathcal{I}_1})$ and $S_2 = (V_2, \{Q_i\}_{i \in \mathcal{I}_2})$ with $V_1 \cap V_2 = \emptyset$, and given vertex $v \in V_1$, the discrete structure formed from S_1 by **substituting S_2 for v** is the discrete structure $S_1[S_2 \to v]$ on vertex set $(V_1 - \{v\}) \cup V_2$ whose relations are $\{R_i'\}_{i \in \mathcal{I}_1} \cup \{Q_i\}_{i \in \mathcal{I}_2}$, where R_i' is formed from R_i by replacing any occurrence of v as a component by *any* $u \in V_2$. The structure $S_1[S_2]$ is formed by successively substituting S_2 for each vertex of S_1; the resulting structure is also called the **lexicographic product** of S_1 and S_2.

We postpone examples of these structures until later, when they will appear in a number of contexts.

2.4 Representations and Models

Discrete structures have been well utilized by mathematicians, computer scientists and others precisely for the fact that they serve as models for many problems, and the capabilities of discrete structures to model have led to the growing interest in their abstract study. Other established areas of mathematics, such as geometry, algebra, analysis and topology can be used to represent discrete structures (and vice versa).

An essential aspect is that the representation should (or could) be the same for isomorphic structures. Let's begin with some of the more useful and common general representations of discrete structures.

2.4.1 Geometric Models

We, as sighted beings, naturally seek a visualization of abstractions. Each type of discrete structure has its own associated geometric representation, usually in the plane $\mathbb{R}^2$. While we shall discuss explicitly a variety of geometric models for each type of discrete structure, the general framework is the same: we take a continuous 1–1 function $\rho : V \to \mathbb{R}^2$ to represent the vertices in the plane, and for each relation R_i of arity n_i (or sometimes, each edge), we use a Jordan arc or curve to "capture" which ordered n_i–tuples belong to R_i (or the edge set) by passing through the images of the associated vertices or by encircling them. For some order relations, height in the plane is also used in part to encode the relation. The images of the vertices are often labeled with the labels of the vertices themselves. Figure 2.1 presents geometric representations of the previously listed examples.

Using geometric models associates with a discrete structure a set of topologies, and thereby opens up discrete structures to a geometric approach.

2.4.2 Algebraic Models

Matrices are a fundamental underpinning for linear algebra, so it is natural to inquire as to whether matrices can be used to model discrete structures. Again, each type of discrete structure has associated with it a variety of encoding schemes by matrices. However, a fundamental one can be defined as follows: for an undirected discrete structure S with vertex set $V = \{v_1, \ldots, v_n\}$ and edge set $E = \{e_1, \ldots, e_m\}$, form an $|V| \times |E|$ **vertex–edge** matrix M such that

$$M_{i,j} = \begin{cases} 1 & \text{if } v_i \in e_j, \\ 0 & \text{otherwise.} \end{cases}$$

For directed structures, we need to alter the definition to take into account the ordering. For example, for a directed graph $D = (V, R)$ with $V = \{v_1, \ldots, v_n\}$ and the ordered pairs in R are $r_1, \ldots, r_m$ we can define

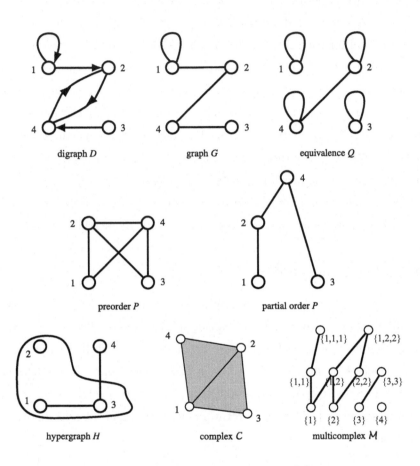

FIGURE 2.1: Some geometric representations of discrete structures

a **vertex–arc matrix**

$$M_{i,j} = \begin{cases} 1 & \text{if } v_i \text{ is the initial vertex of } r_j, \\ -1 & \text{if } v_i \text{ is the terminal vertex of } r_j, \\ 0 & \text{otherwise.)} \end{cases}$$

The definition of the vertex–edge matrix M depends on the ordering of both V and E, but differs only by a permutation of rows and columns (that is, for any two such matrices M_1 and M_2, there are permutation matrices P and Q such that $M_2 = PM_1Q$). If one wishes, one can associate square matrices with the discrete structure in a number of ways, say by considering the matrices M^TM or MM^T.

Representing discrete structures by matrices suggests investigating the connection between algebraic properties and parameters of the matrix (such as rank, determinant, existence of an inverse, eigenvalues, etc.) and combinatorial ones of the original discrete structure.

Example: Consider the graph $F = (V, E)$, where $V = \{1, 2, 3, 4\}$ and $E = \{\{1, 2\}, \{2, 4\}, \{3, 4\}\}$ (this is the underlying loopless graph for the graph G shown in Figure 2.1). A vertex-edge matrix of F is

$$M = \begin{pmatrix} 1 & 0 & 0 \\ 1 & 1 & 0 \\ 0 & 0 & 1 \\ 0 & 1 & 1 \end{pmatrix},$$

and so

$$MM^T = \begin{pmatrix} 1 & 1 & 0 & 0 \\ 1 & 2 & 0 & 1 \\ 0 & 0 & 1 & 1 \\ 0 & 1 & 1 & 2 \end{pmatrix} \quad \text{and} \quad M^TM = \begin{pmatrix} 2 & 1 & 0 \\ 1 & 2 & 1 \\ 0,2,0 & 1 & 2 \end{pmatrix}.$$

It is not hard to check that MM^T has eigenvalues $0, 2, 2 - \sqrt{2}, 2 + \sqrt{2}$ while M^TM has eigenvalues $2, 2 - \sqrt{2}, 2 + \sqrt{2}$ (see Exercise 2.11), and hence both matrices are diagonalizable. Both matrices have rank 3 (why?) and so MM^T has no inverse while M^TM does. △

There are other ways to encode a discrete structure algebraically. For example, for a hypergraph H with vertex and edge sets $V = \{v_1, \ldots, v_n\}$ and E respectively, one can, over any field $\mathbf{k}$, take the vector space $\mathbf{k}(V)$ over $\mathbf{k}$ with basis V. Representing the edge e of H by

$$\mathbf{e} = \sum_{v \in e} v$$

one can consider the subspace generated by $\{\mathbf{e} : e \in E\}$ as an algebraic model for H, and ask questions about its dimension, etc.

As another example, a multicomplex $\mathcal{M}$ on set $V = \{v_1, \ldots, v_n\}$ can be embedded in the polynomial ring $\mathbf{k}[v_1, \ldots, v_n]$ by representing multiset m by

$$\mathbf{m} = \prod_{v \in m} v.$$

What is so nice about this representation of a multicomplex by monomials is that multiset containment is converted into divisibility of the associated monomials. It turns out that this representation will be very useful in the deeper connections between combinatorics and commutative algebra.

2.4.3   Logical Models

First-order logic contains the usual connectives $\neg, \wedge, \vee, \rightarrow$ ("not", "and", "or" and "implies", respectively), the quantifiers $\forall$ and $\exists$, variables, predicates $S(x_1, \ldots, x_m)$ (for some positive integer m), the binary predicate equality $=$, and constants. A **first-order theory** is simply a set of sentences from first-order logic. Discrete structures can be modeled directly in first-order logic. For an l–ary relation R, one can take an l–ary **predicate** $P(x_1, \ldots, x_l)$ and create a first-order theory of the structure.

Example: The first-order theory of partial orders (with predicate P) is

$$PO = \{(\forall x)(P(x, x)), (\forall x)(\forall y)((P(x, y) \wedge P(y, x)) \rightarrow x = y),$$
$$(\forall x)(\forall y)(\forall z)(((P(x, y) \wedge P(y, z)) \rightarrow P(x, z))\};$$

any **(logical) model** of the theory PO (with a domain of some set V and some binary relation R interpreting the binary predicate P) will be a partial order. $\triangle$

The connection to logic is a deep and strong one. One can use the deductive power of first-order logic to deduce theorems on discrete structures. Important results from first order logic that we will make use of are mentioned in the appendices.

2.4.4 Probabilistic Models

Often one wishes to enumerate all discrete structures on a given finite set V, to approximate the proportion of such structures with a given property, or to prove the existence of a single structure with a given property. In such cases, it has been helpful to create a randomized version of the structure (or an important subclass of such structures). One simple way to do so is to assign to each of the finite number of such discrete structures the same probability. In other cases, it is more helpful to generate out the discrete structures in some other way, so that the mechanism for generation is simplified and more amenable to calculation, while the individual structures may have varying probabilities. Due to the nature of the properties of various discrete structures, it may be difficult to generate them in any systematic way at random.

As an example, suppose we let P_k be the graph property that for any two disjoint subsets U and W of the vertex V of a graph G, each of size at most k, there is a vertex $x \in V - (U \cup W)$ such that v is joined to every vertex of U and no vertex of W. What we would like to do is find graphs with property P_k. This is not difficult for $k = 1$ (see Figure 2.2), but for $k \geq 2$ this is nontrivial. But probability comes to the rescue to show that almost *every* graph is an example! Let us make every simple graph of order n (that is, on the set $\{1, \ldots, n\}$) equally probable, each with probability $2^{-\binom{n}{2}}$). Equivalently, we think of independently flipping a fair coin for pair of point x and y, and if the coin comes up heads, we put in the edge $\{x, y\}$, and otherwise we don't.

Theorem 2.1 *Let k be a fixed positive integer. Then*

$$\lim_{n \to \infty} Prob(G \ has \ property \ P_k) = 1.$$

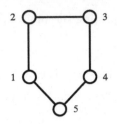

FIGURE 2.2: A graph with property P_1

Proof: Let n be a large positive integer greater than $2k$. For subsets U and W of $V = \{1, \ldots, n\}$ (the vertex set) of size k each, and vertex $z \in V - (U \cup W)$, let $A_{z,U,W}$ be the event that z is joined to all vertices of U and to no vertex of W. Then clearly

$$\text{Prob}(A_{z,U,W}) = \left(\frac{1}{2}\right)^k \left(\frac{1}{2}\right)^k = 2^{-2k},$$

as all of the k edges between z and U need to be present and all of the k edges from z to W must be missing. It follows that if $E_{z,U,W}$ is the complementary event to $A_{z,U,W}$, that is, the probability that z is not joined to all vertices of U and to no vertex of W, then

$$\text{Prob}(E_{z,U,W}) = 1 - \text{Prob}(A_{z,U,W}) = 1 - 2^{-2k}.$$

If we now let $E_{U,W}$ denote the event that *no* vertex $z \in V - (U \cup W)$ is joined to all vertices of U and to no vertex of W, then

$$E_{U,W} = \bigcap_{z \in V - (U \cup W)} E_{z,U,W}.$$

Moreover, for given u and W, the events $\{E_{z,U,W} : z \in V - (U \cup W)\}$ are independent of one another as the sets of possible edges from different z's to

U and W are disjoint. From basic probability, we have

$$\text{Prob}(E_{U,W}) = \text{Prob}\left(\bigcap_{z \in V-(U \cup W)} E_{z,U,W}\right)$$

$$= \prod_{z \in V-(U \cup W)} \text{Prob}(E_{z,U,W})$$

$$= \prod_{z \in V-(U \cup W)} \left(1 - 2^{-2k}\right)$$

$$= \left(1 - 2^{-2k}\right)^{n-2k}$$

as there are $n - 2k$ choices for z.

Now if a graph fails to satisfy property P_k, then there must be some sets U and W such that *no* vertex z is joined to all vertices of U and to no vertex of W, that is, some $E_{U,W}$ must hold. So we have

$$\text{Prob}(G \text{ does not have property } P_k) \leq \text{Prob}\left(\bigcup_{U,W} E_{U,W}\right)$$

$$\leq \sum_{U,W} \text{Prob}(E_{U,W})$$

$$= \sum_{U,W} \left(1 - 2^{-2k}\right)^{n-2k}$$

$$= \binom{n}{k}\binom{n-k}{k}\left(1 - 2^{-2k}\right)^{n-2k}$$

$$\leq n^{2k}\left(1 - 2^{-2k}\right)^{n-2k}.$$

Consider what happens to $n^{2k}\left(1 - 2^{-2k}\right)^{n-2k}$ as n gets large. As k is fixed, $\left(1 - 2^{-2k}\right)$ is a fixed real number x between 0 and 1. The natural logarithm of

$$n^{2k}\left(1 - 2^{-2k}\right)^{n-2k} = n^{2k}x^{n-2k}$$

is

$$2k \ln n + (n - 2k)\ln x$$

which tends to $-\infty$ as $n \to \infty$, since $\ln x$ is negative. It follows that

$$\lim_{n \to \infty} \text{Prob}(G \text{ does not have property } P_k) = 0,$$

which implies that

$$\lim_{n \to \infty} \text{Prob}(G \text{ has property } P_k) = 1.$$

∎

The end result is that for large n, almost every graph on n vertices has property P_k, even though it is hard to come up with any explicit examples. The story is that the discrete objects we tend to think of are anything but random.

Exercises

Exercise 2.1 *Suppose $\preceq$ is a reflexive relation on V. Show that $P = (V, \preceq)$ is a poset if and only if $\prec = \preceq - \{(x, x) : x \in V\}$ is irreflexive and transitive.*

Exercise 2.2 *Draw a diagram indicating the containment relation among the various types of discrete structures: digraphs, graphs, equivalences, preorders, partial orders, hypergraphs, simplicial complexes and multicomplexes.*

Exercise 2.3 *How many possible isomorphisms are there between discrete structures (V, R) and (U, S), if $|V| = |U| = n$?*

Exercise 2.4 *Prove that the each of the induced substructures, spanning substructures and partial substructures of a discrete structure are partial orders under the binary relation $\subseteq$ (for both vertex sets and relations).*

Exercise 2.5 *How many induced substructures and spanning substructures of a discrete structure $D = (V, R)$ are there, if $|V| = n$ and $|R| = m$?*

Exercise 2.6 *Prove that if two digraphs are isomorphic then one is a partial order if and only if the other is.*

Exercise 2.7 *Suppose that $D_1 = (V, R)$ and $D_2 = (U, Q)$ are two disjoint (finite) discrete structures, with $v = |V|$, $u = |U|$, $r = |R|$, $q = |Q|$, R an n-ary relation and Q an m-ary relation. What is the cardinality of the relation set for the join of D_1 and D_2? What is the cardinality of the relation set for $D_1[D_2]$?*

Exercise 2.8 *Give an example of graphs G_1 and G_2 such that $G_1[G_2]$ is not isomorphic to $G_2[G_1]$.*

Exercise 2.9 *For the digraph shown below, which of the following is it: a graph, an equivalence relation, a partial order or a preorder?*

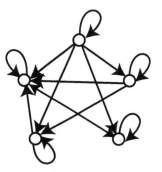

Exercise 2.10 *Prove that both $M^T M$ and $M M^T$ are positive semidefinite (see the appendix to recall the definition of a positive definite matrix).*

Exercise 2.11 *Prove that both $M^T M$ and $M M^T$ have the same nonzero eigenvalues.*

Exercise 2.12 *Suppose that $D = (V, A)$ is a directed graph and M is the vertex–arc matrix. Prove that at least one of $M^T M$ and $M M^T$ is singular.*

Exercise 2.13 *Suppose that $G = (V, E)$ is an undirected graph with vertex–edge matrix M. Let $A = M M^T$. Explain what the entries in A mean in terms of the graph.*

Exercise 2.14 *Explain how to use the vertex–edge matrix of H to calculate the dimension of the subspace generated by $\{\mathbf{e} : e \in E\}$ as defined in (2.1).*

Exercise 2.15 *Ehrenfeucht [37] and Fraisse [44] proved the following general theorem connecting the first-order theory of discrete structures and games. Consider the first-order theory of all discrete structures with a given (finite) number of relations of fixed arity. One can define the* quantifier depth *in a natural way recursively for any sentence of the theory (it is the maximum over $\wedge$, $\rightarrow$ and $\vee$, the same under negation, and one more if an additional $\forall$ or*

∃ is tacked on in front). Consider any two relevant discrete structures (the underlying vertex sets may be finite or infinite). Player I chooses a vertex in one discrete structure; player II responds by choosing a vertex in the other structure. Player I then proceeds again, choosing a vertex in either structure, while player II responds in the other structure. The game proceeds for k plays ($k \geq 1$). Let the vertices of D_1 chosen in order be $x_1, \ldots, x_k$ and let the vertices of D_2 chosen in order be $y_1, \ldots, y_k$. Then player II wins if and only if the map ρ that sends each x_i to y_i ($i = 1, \ldots, k$) is an isomorphism of the induced substructures. The amazing results in [37, 44] show that player II has a winning strategy if and only if the structures D_1 and D_2 are indistinguishable by sentences of quantifier depth at most k in the first-order theory.

Prove that if D_1 is a graph with a universal vertex (that is, a vertex joined to all other vertices) and D_2 is a graph without a universal vertex, then player II has a winning strategy in 2 steps. Can you find a sentence of quantifier depth 2 that expresses the fact that a graph has a universal vertex?

Exercise 2.16 *Ehrenfeucht–Fraisse games (see the previous exercise) can be used to prove the non–expressibility of certain properties in first-order logic. By considering plays of the game with an infinite two–way path $G_1 = P_\infty$ and the disjoint union of two such paths $G_2 = P_\infty \cup P_\infty$, show that connectedness cannot be expressed as a first-order sentence.*

Exercise 2.17 *Find infinitely many graphs with property P_1.*

Exercise 2.18 *Suppose for fixed n we consider as our sample space loopless graphs on $\{1, \ldots, n\}$. If all such graphs are equally likely, what is the probability of any specific graph of order n occurring?*

Exercise 2.19 *We can make all k–uniform hypergraphs on $\{1, \ldots, n\}$ into a sample space by choosing each k-subset independently with probability p ($p \in [0, 1]$ fixed). What is the expected number of edges in a random k–uniform hypergraph? What is the variance in the number of edges?*

Chapter 3

Graphs and Directed Graphs

We begin by discussing some theory of graphs and digraphs. Before we do, we shall remind ourselves of some standard notation. The **complete graph** K_n on V is any simple graph of order n with all $\binom{n}{2}$ edges. The **empty graph** E_n is any graph of order n with no edges. The **path** P_n of order n has vertex set $[n]$ with edges $\{i, i+1\}$ for $i = 0, \ldots, n-2$. The **cycle** C_n of order n has vertex set $[n]$ with edges $\{i, i+1\}$ for $i = 1, \ldots, n$, with addition modulo n (a cycle of order 1 is a loop, while a cycle of order 2 consists of two parallel edges). The complete k–**partite** graph $K_{l_1, l_2, \ldots, l_k}$ has vertex set $V = V_1 \cup \cdots \cup V_k$, where the V_i's are disjoint and $|V_i| = l_i$, and edge set $\{uv : u \in V_i, \ v \in V_j, \ i \neq j\}$. In the case $k = 2$, we say **bipartite** rather than 2–partite. The **stars** are the graphs $K_{1,l}$. A **forest** is a graph that contains no cycles, and a **tree** is a connected forest.

There are no more important representations of graphs and digraphs than the standard geometric ones. A geometric representation of a directed graph $D = (V, A)$ is a collection of continuous functions $\rho : V \to \mathbb{R}^2$ and $\phi_a : [0, 1] \to \mathbb{R}^2$, for all $a \in A$, such that ρ is 1–1, and for each $a = (u, v) \in A$, ϕ_a is a Jordan arc or curve (i.e. ϕ_a restricted to $[0, 1)$ is 1–1), $\phi_a(0) = \rho(u)$, $\phi_a(0) = \rho(v)$

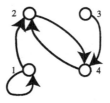

FIGURE 3.1: A geometric representation of a digraph

and $\phi_a((0,1)) \cap \{\rho(w) : w \in V\} = \emptyset$. If moreover $\phi_a((0,1)) \cap \phi_b((0,1)) = \emptyset$ for distinct arcs a and b, then the representation is **proper**). The image of the maps is what is usually drawn "as" the directed graph, with arrows on each arc indicating the direction (i.e. if $a = (u, v)$, then the arrow on $\phi_a([0,1])$ points towards $\phi_a(v)$).

Example: Consider $V = \{1, 2, 3, 4\}$ and $A = \{(1,1), (1,2), (2,4), (4,2), (3,4)\}$. A geometric representation of directed graph $D = (V, A)$ is shown in Figure 3.1. (Here, and elsewhere, the points representing vertices are enlarged for viewing purposes.)   △

A geometric representation of a graph $G = (V, E)$ is a collection of continuous functions $\rho : V \to \mathbb{R}^2$ and $\phi_e : [0,1] \to \mathbb{R}^2$, for all $e \in E$, such that ρ is 1–1, and for each $e = \{u, v\} \in E$, ϕ_e is a Jordan arc or curve, $\phi_e(0) = \rho(u)$ and $\phi_e(0) = \rho(v)$ and $\phi_e((0,1)) \cap \{\rho(w) : w \in V\} = \emptyset$ for all edges s. If moreover $\phi_e((0,1)) \cap \phi_f((0,1)) = \emptyset$ for distinct edges e and f, then the representation is **proper**). The image of the maps is what is usually drawn "as" the graph.

Example: Consider $V = \{1, 2, 3, 4\}$ and $R = \{(1,1), (1,2), (2,1), (2,4), (4,2), (3,4), (4,3)\}$. A geometric representation of graph $G = (V, R)$ is shown in Figure 3.1.   △

Example: Geometric representations of some other graphs are shown in Figure 3.2.   △

We sometimes represent graphs and digraphs in higher order Euclidean spaces. For example, we can represent any graph G or digraph D in $\mathbb{R}^3$ by placing the vertices along the z–axis, and drawing each edge a Jordan arc on a different plane through the z–axis; such a drawing ensures that no two edges meet except at its endpoints.

We now turn to algebraic representations. Besides the vertex–edge adjacency matrix, one often represents a digraph (and graph) by its **adjacency**

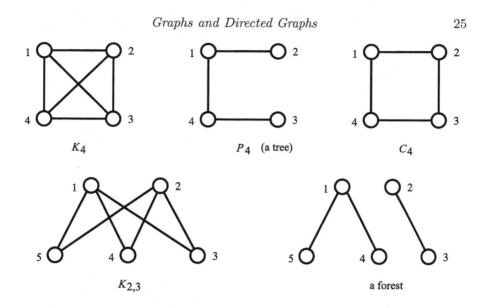

FIGURE 3.2: Geometric representations of some graphs

matrix. Given a digraph $D = (V, A')$ on vertex set $V = \{v_1, \ldots, v_n\}$, the associated $n \times n$ adjacency matrix is A, where

$$A_{i,j} = \begin{cases} 1 & \text{if } (v_i, v_j) \in A' \\ 0 & \text{otherwise.} \end{cases}$$

For a graph G with vertex set $V = \{v_1, \ldots, v_n\}$ and edge set E, the associated $n \times n$ adjacency matrix is A, where

$$A_{i,j} = \begin{cases} 1 & \text{if } \{v_i, v_j\} \in E \\ 0 & \text{otherwise.} \end{cases}$$

(Loops usually count "twice" on the diagonal; for multidigraphs (resp. multigraphs) we extend the definition for $A_{i,j}$ to count the number of times a given ordered pair (unordered pair) is an arc (edge).)

Example: The adjacency matrix of the complete graph K_n is $J_n - I_n$, where I_n is the $n \times n$ identity matrix and J_n is the $n \times n$ consisting of all 1s. An adjacency matrix for the digraph of Figure 3.4 is

$$A = \begin{pmatrix} 1 & 1 & 0 & 0 \\ 0 & 0 & 0 & 1 \\ 0 & 0 & 0 & 1 \\ 0 & 1 & 0 & 0 \end{pmatrix}.$$

△

If D is a graph, then A is symmetric. It is not hard to verify (by induction) the following important result.

Proposition 3.1 *For directed graph D on vertex set $\{1, \ldots, n\}$ and integer $k \geq 1$, the (i, j)–th entry of A^k is the number of i–j walks in D of length k.*

Proof: For $k = 1$, the result is trivial, as the number of walks of length 1 from i to j is the number of arcs (i, j) in the directed graph D, and this is by definition the (i, j)–th entry of $A = A^1$. So assume now that the result holds for k, that is, for i and j, the (i, j)–th entry of A^k is the number of i–j walks in D of length k. We need to show that the result holds for $k + 1$ as well. So consider any i and j. We can enumerate all i–l walks of length $k + 1$ as a walk of length k from i to some vertex l, followed by an arc from l to j. By assumption, the number of the former is A^k_{il}, the (i, l)–th entry of A^k, while the number of (l, j) arcs is A_{lj}. It follows that the number of i–j walks of length $k + 1$ is

$$\sum_{l=1}^{n} A^k_{il} \cdot A_{lj}$$

and by the definition of matrix product, this is the same as A^{k+1}_{ij}, the (i, j)–th entry of A^{k+1}. This completes the induction. ∎

A related matrix for loopless multigraphs $G = (V, E)$ of order n is the $n \times n$ **Laplacian** L, which is defined as

$$L_{i,j} = \begin{cases} \deg_G(v_i) & \text{if } i = j \\ -\mu(v_i, v_j) & \text{if } i \neq j \end{cases}$$

where for $i \neq j$, $\mu(v_i, v_j)$ is the number of times $\{v_i, v_j\}$ is an edge of G. The

Laplacian of a multigraph has the property that it is positive semidefinite, and hence has nonnegative real eigenvalues.

One important theorem about Laplacians is the following. Its proof can be found in many texts on algebraic graph theory (c.f. [9, pg. 39]).

Theorem 3.2 (The Matrix Tree Theorem) *Let L be the Laplacian of a loopless multigraph. Then any cofactor of L is equal to the number of spanning trees of G (two spanning trees are considered different if they differ in at least one edge).* ■

Example: Let $G = K_4 - e$ (the graph formed from the complete graph of order 4 by removing an edge) be the graph in Figure 3.3. Its Laplacian (under the vertices listing $1, 2, 3, 4$) is given by

$$L = \begin{pmatrix} 2 & -1 & -1 & 0 \\ -1 & 3 & -1 & -1 \\ -1 & -1 & 3 & -1 \\ 0 & -1 & -1 & 2 \end{pmatrix}.$$

The $(1,1)$–cofactor of L is equal to the determinant of the matrix

$$\begin{pmatrix} 3 & -1 & -1 \\ -1 & 3 & -1 \\ -1 & -1 & 2 \end{pmatrix}$$

which is 8. One can verify that indeed G has exactly 8 spanning trees (there are 4 spanning trees not containing the edge $\{2,3\}$, and there are 4 spanning trees that do contain the edge $\{2,3\}$). △

There is a vast list of standard definitions for graphs and digraphs; we have seen some and shall only list a few more of the important ones we shall need. The **underlying graph** of a digraph $D = (V, A)$ is the graph G on V such that $\{x, y\}$ is an edge of G if and only if $(x, y) \in A$ or $(y, x) \in A$. Following our discussion of general discrete structures in Chapter 2, for vertices u and v of digraph D (resp. graph G) a u–v **walk** of length k is an alternating sequence

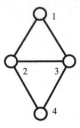

FIGURE 3.3: The graph $K_4 - e$

$u = w_0, e_0, w_1, e_1, \ldots, e_k, w_{k+1} = v$, where for all $l \in \{1, \ldots, k\}$, e_i is an arc (edge) of D (G) from w_{i-1} to w_i. A **circuit** in a digraph or graph is a closed walk (i.e. one that starts and ends at the same vertex) with distinct edges; a **cycle** is a circuit without repeated vertices (except that the first and last vertices are the same).

A digraph is **strongly connected** if there are u–v and v–u walks for all vertices u and v (if the digraph is in fact a graph, we merely say G is **connected**). A digraph is **weakly connected** if and only if its underlying graph is connected.

An **independent set** I of a graph G is a subset of vertices that induce a subgraph that contains no edges; a **clique** of a simple graph G is a subset of vertices that induce a subgraph that has every pair of points joined by an edge (we often identify both independent sets and cliques with the subgraphs they induce). The **independence number** $\beta(G)$ is the order of the largest independent set in G, while the **clique number** $\omega(G)$ is the order of the largest clique in G.

Example: The underlying graph of the digraph D in Figure 3.4 is the graph G of the same figure. The graph G is connected (and hence D is weakly connected), while the digraph D is not strongly connected (though it is weakly connected). $\{1, 3\}$ is an independent set of G, and $\{2, 4\}$ is a clique of G. G has independence number 2 a clique number 2. $\triangle$

It is surprising how often one can use a simple result. Here is one we shall use later; we'll leave the proof to you!

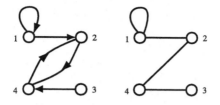

FIGURE 3.4: A digraph and its underlying graph

Theorem 3.3 *If G is a finite graph with all vertices of degree 2 then G decomposes uniquely into cycles. If D is a finite directed graph with all indegrees and outdegrees equal to 1 then D decomposes uniquely into (directed) cycles.*
∎

3.1 Graphs and Directed Graphs as Models

Graphs and directed graphs form the basis of much of computer science, as many applications deal with the relationship between pairs of items. As well, graphs and directed graphs are also used as models within pure mathematics itself. Let's look at some examples.

3.1.1 Graph Colourings

A **(proper vertex) k–colouring** of a graph $G = (V, E)$ is a function $f : V \to \{1, \ldots, k\}$ such that for any edge $\{u, v\}$ of G, $f(u) \neq f(v)$. Note that each **colour class** (i.e. the set of vertices receiving a fixed colour) induces an independent set of G. The **chromatic number**, $\chi(G)$, is the least k for which G has a k–colouring. Graph G is $mathbf{k}$**–colourable** if G has a k–colouring.

A graph G is **vertex k–critical** if $\chi(G) = k$ but $\chi(G - v) < k$ for all vertices v of G, while G is **(edge) k–critical** if $\chi(G) = k$, G is connected and $\chi(G - e) < k$ for all edges e of G.

Example: K_n is n–critical as it is obviously n–chromatic (as every vertex needs

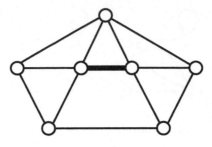

FIGURE 3.5: A vertex 4–critical graph that is not 4–critical

a different colour) and connected, while the removal of any edge allows it to be coloured with one fewer colour. The odd cycles C_{2n+1} are 3–critical, as they are connected and not 2–colourable, but the removal of any edge leaves a path, which is 2–colourable.   △

Example: The graph shown in Figure 3.5 is vertex 4–critical, but not 4–critical, as removing the edge shown in bold still leaves a 4–chromatic graph (this is the smallest graph that is vertex critical but not critical.   △

Suppose we want to schedule examinations at a university. There are a variety of courses whose examinations must be scheduled within, say, 10 days, each having 3 examination periods available. There is a constraint; if a student takes two courses, the corresponding examinations should be scheduled in different time slots. If we form a graph whose vertex set is the set of examinations and whose edges correspond to examinations that cannot be scheduled at the same time, then a schedule corresponds precisely to a colouring of the graph, and vice versa. Thus graph colouring is a model for examination scheduling. (Of course, there may be constraints as to which time slots an examination may or may not be scheduled, but these translate into conditions on the nature of the permissible colouring allowed.)

Graph colourings have also arisen in pure mathematics. For example, a problem of interest a while ago was to determine how few sets are needed to partition the plane $\mathbb{R}^2$ so that no two points at distance 1 are in the same set. This is indeed a graph colouring problem, as, by forming a graph $\mathbb{G}^2$ on vertex set $\mathbb{R}^2$ whose edges correspond to points at distance 1, we are seeking the chromatic number of this graph. It is known that $\chi(\mathbb{G}^2) \leq 7$ (Exercise 3.13) and that $\chi(\mathbb{G}^2) \geq 4$, as there is a 4-chromatic finite induced subgraph of $\mathbb{G}^2$ (Exercise 3.14). Note that we are colouring an infinite graph (indeed an uncountable one!), but we shall see in Section 3.2.4 that, to some extent, we need not consider the whole graph, but only finite pieces of it.

3.1.2   Reliability

Here is another applied problem. Suppose we have a network (i.e. multigraph) G for which the nodes (i.e. vertices) are always operational (i.e. working) but for which each edge e is independently operational with probability p_e. An important consideration in such a network is whether, with high likelihood, information can travel from any one vertex to another. The **(all terminal) reliability** of G is therefore defined to be the probability that the operational edges form a spanning connected subgraph of G (such a subgraph is often called an **operational state** of the graph). The most amenable case is that in which all of the edge probabilities are identical, say p, and we denote the subsequent reliability by $\mathrm{Rel}(G, p)$ (the more general case, with edge e having probability p_e of being operational, is denoted by $\mathrm{Rel}(G, \{p_e : e \in E\})$). Loops have no effect on reliability (though multiple edges certainly do), so we often (but not always) assume our multigraphs to be loopless.

Example: The reliability of a tree T_n of order n is p^{n-1} as all $n - 1$ edges must be operational. The reliability of the multigraph of order 2 with the two vertices joined by k edges has reliability $1 - (1 - p)^k$, as the multigraph is operational as long as at least one edge is operational. $\triangle$

How does one calculate reliabilities? On one hand, you could simply list all possible connected subgraphs of G, partitioned into classes according to how many edges they have, and add up the probability of each subgraph occurring. If G has order n and size m, and F is a connected spanning subgraph of size $m - i$, then F has probability $p^{m-i}(1 - p)^i$ of occurring, as each edge of F must be operational and the edges of G not in F must be non-operational.

Thus adding up over all choices for F, we would deduce that $\text{Rel}(G, p)$ has the form (its F–**form**)

$$\text{Rel}(G, p) = \sum_i F_i p^{m-i}(1 - p)^i$$

where each F_i is a nonnegative integer (our reasoning for choosing to enumerate the list by how many edges are *non-operational* rather than by how many edges are operational will only be clear later on). This formulation shows that reliability is indeed a polynomial in p.

Example: The reliability of a cycle of order n is $p^n + np^{n-1}(1 - p)$, as the operational states have either 0 or 1 edge non-operational (and any one edge can be non-operational); here $F_0 = 1$, $F_1 = n$, and $F_i = 0$ for $i \geq 2$. We have seen that the reliability of the multigraph of order 2 with the two vertices joined by k edges has reliability $1 - (1 - p)^k$, which can be rewritten as $\sum_{i=0}^{k-1} \binom{k}{i}(1 - p)^i p^{k-i}$, so that $F_i = \binom{k}{i}$ if $i = 0, \ldots, k - 1$, and 0 otherwise. $\triangle$

Example: The reliability of K_4 is $p^6 + 6p^5(1 - p) + 15p^4(1 - p)^3 + 16p^3(1 - p)^3$, as any subgraph on at least 4 edges is operational, and the subgraphs with 3 edges that are operational are those that are a spanning tree of K_4 (of which there are $4^2 = 16$); there are no operational states with 2 or fewer edges, as we need to contain at least a spanning tree. $\triangle$

For an edge $e = \{x, y\}$ of G, let the **contraction** $G \bullet e$ of e in G be the graph on $V - \{x, y\} \cup \{z\}$, where z is a new vertex not in G, with edge multiset being those edges in the induced subgraph on $V - \{x, y\}$ together with all edges $\{\{z, w\} : w \in V - \{x, y\}, \{x, w\} \in E \text{ or } \{y, w\} \in E\}$. A simple argument from elementary probability theory gives the following.

Theorem 3.4 (The Reliability Factor Theorem) *Let G be any multigraph. Then for any edge e of G,*

$$\text{Rel}(G, p) = p\text{Rel}(G \bullet e, p) + (1 - p)\text{Rel}(G - e, p).$$

Proof: Consider the sample space Ω whose elements are the spanning subgraphs of G, with the probability of H being $p^{|E(H)|}(1-p)^{|E(G)-E(H)|}$. Let ϕ be the event that H is connected, so that the reliability of G is the probability of ϕ occurring. Now the probability that ϕ occurs, given that edge e is operational, is $\mathrm{Rel}(G \bullet e, p)$, as if e is operational, then the two endpoints of e can always communicate with each other, and hence we can contract them into a single (always operational) vertex. The probability that ϕ occurs, given that edge e is non-operational, is $\mathrm{Rel}(G - e, p)$, as if e is non-operational, we might as well delete it from the multigraph. By Bayes' Theorem, we have

$$
\begin{aligned}
\mathrm{Prob}(\phi) \;=\; & \mathrm{Prob}(e \text{ is operational})\mathrm{Prob}(\phi|e \text{ is operational}) + \\
& \mathrm{Prob}(e \text{ is non-operational})\mathrm{Prob}(\phi|e \text{ is non-operational})
\end{aligned}
$$

so we conclude that

$$
\mathrm{Rel}(G,p) \;=\; p\mathrm{Rel}(G \bullet e, p) + (1-p)\mathrm{Rel}(G - e, p).
$$

■

Example: The calculation of the reliability of K_4 using the Factor Theorem is shown in Figure 3.6 (the reliability of a graph is abbreviated by its geometric representation). The polynomials are calculated from the bottom up. △

The Factor Theorem shows again that $\mathrm{Rel}(K_4, p)$ is a polynomial in p. We can, in fact, show more (see Exercise 3.21):

Theorem 3.5 *Let G be a connected multigraph of order n and size m. Then $Rel(G, p)$ is a polynomial in p of degree m with integer coefficients that alternate in sign. Moreover, the smallest power of p occurring is $n - 1$, and the coefficient of this term is the number of spanning trees of G.*

Reliability is therefore an algebraic or analytical model, saying something about the "connectedness" of the multigraph. We'll delve more into this connection shortly.

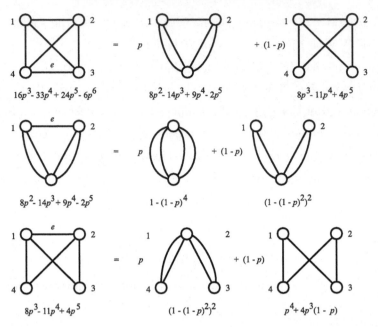

FIGURE 3.6: Rel(K_4, p) as calculated by the Factor Theorem

3.1.3 Proofs in Matrix Theory

While it is well known that matrices can model directed graphs, it may be surprising how well directed graphs can be used to model matrices. We're going to prove the famous and important Cayley–Hamilton Theorem from linear algebra using only discrete mathematics (the argument is due to Straubing [93] – see also Zeilberger [106]). And what is more, we shall make no reference to the underlying field – the argument will show that the Cayley–Hamilton Theorem is true <u>formally</u>, rather than merely algebraically! Recall what the Cayley–Hamilton Theorem states:

The Cayley–Hamilton Theorem If A is an $n \times n$ matrix with characteristic polynomial $c(\lambda) = \det(\lambda I - A)$, then $c(A) = 0$, where 0 denotes the $n \times n$ 0 matrix.

By expanding $\det(\lambda I - A)$ (using the multilinearity of determinants), one can see that the coefficient of λ^{n-k} is the sum of all **principal k × k minors** of

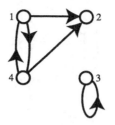

FIGURE 3.7: A digraph D

$-A$ (a principal $k \times k$ minor of a matrix is the determinant of the submatrix formed by taking k rows and the corresponding k columns). Thus we shall prove that

$$
\begin{aligned}
0 = \ & A^n + (-a_{11} - a_{22} - \cdots - a_{nn})A^{n-1} + \cdots + \qquad (3.1) \\
& \left(\sum \text{principal } k \times k \text{ minors of } -A\right)A^{n-k} + \cdots + \det(-A).
\end{aligned}
$$

Consider a digraph on vertex set $[n] = \{1, 2, \ldots, n\}$. We extend our notion of a digraph to one of a **weighted digraph**, where each edge e has associated to it a value or **weight** $w(e)$. Here we will attach the weight of $a_{i,j}$ to an edge (i, j) (loops are permitted). This is how we model the matrix A by a digraph.

Now for any weighted digraph D, we define the weight of D to be

$$
w(D) = \prod_{e \in A(D)} w(e).
$$

Moreover, for any family $\mathcal{F}$ of digraphs on $[n]$, we define the weight of the family to be

$$
w(\mathcal{F}) = \sum_{D \in \mathcal{F}} w(D).
$$

Example: The weight of the digraph D shown in Figure 3.7 is $w(D) = a_{1,2}a_{3,3}a_{1,4}a_{4,1}a_{4,2}$. $\triangle$

We need another definition that defines an alternate weight for a digraph that is the disjoint union of (directed) cycles. Suppose D, a digraph on $[n]$, is the disjoint union of cycles $C_1, C_2, \ldots, C_l$. Then we define

$$
w'(D) = (-1)^l \prod_{i=1}^{l} w(C_i).
$$

FIGURE 3.8: A digraph that is a disjoint union of cycles

We extend w' to families of digraphs on $[n]$ as above, that is

$$w'(\mathcal{F}) = \sum_{D \in \mathcal{F}} w'(D).$$

Example: For the digraph D shown in Figure 3.8 (which is a disjoint union of cycles), $w'(D) = -a_{1,4}a_{4,1}a_{2,2}a_{3,3}$.   △

Let Perm_n be the set of all permutations on $[n]$. For any $\pi \in \mathrm{Perm}_n$, we define the **permutation digraph** D_π as the one on $[n]$ with arcs $(i, \pi(i))$. Any permutation digraph has indegrees and outdegrees equal to 1, so by Theorem 3.3, we see that D_π decomposes uniquely into cycles, the same cycles as the permutation decomposes into. As the sign of a permutation cycle is one fewer than its length and the sign of a permutation is the product of the signs of its cycles, we see that

$$w'(D_\pi) \quad = \quad \mathrm{sign}(\pi)(-a_{1\pi(1)})\ldots(-a_{n\pi(n)}) \qquad (3.2)$$

so that

$$w'(\mathrm{Perm}_n) = \det(-A).$$

A similar argument show that if $\mathcal{F}_i$ is the family of all digraphs on $[n]$ that are disjoint union of cycles covering exactly k vertices, then

$$w'(\mathcal{F}_i) \quad = \quad \sum(\text{principal } k \times k \text{ minors of } -A). \qquad (3.3)$$

We now fix i and j in $[n]$. We define the class $\mathcal{F}_{i,j}$ as the set of all ordered pairs (P, C) where P is an i–j walk, C is a disjoint union of (cycles (both on

$[n]$) and the total number of edges in P and C is n. For any $(P,C) \in \mathcal{F}_{i,j}$ we define its weight to be

$$\text{wt}(P,C) = w(P)w'(C)$$

and extend this to $\mathcal{F}_{i,j}$ by

$$\text{wt}(\mathcal{F}_{i,j}) = \sum_{(P,C)\in\mathcal{F}_{i,j}} \text{wt}(P,C).$$

It is not hard to verify (from Proposition 3.1 and (3.3)) that $\text{wt}(\mathcal{F}_{i,j})$ is equal to the (i,j) entry of the right-hand side of (3.1). Thus it remains to show that $\text{wt}(\mathcal{F}_{i,j}) = 0$.

How do we do this? We shall show that there is a map $\alpha : \mathcal{F}_{i,j} \to \mathcal{F}_{i,j}$ such that α^2 is the identity map on $\mathcal{F}_{i,j}$ and α **reverses signs**, that is, for any $(P,C) \in \mathcal{F}_{i,j}$, $\text{wt}(\alpha(P,C)) = -\text{wt}(P,C)$. Thus α pairs the elements of $\mathcal{F}_{i,j}$ into pairs whose weights sum to 0, so that the weight of $\mathcal{F}_{i,j}$ is 0 as well (α is thus called a **"killer" involution**).

The definition of α is rather easy. For any $(P,C) \in \mathcal{F}_{i,j}$, we walk along P from i until one of two things happens:

- we return to a vertex of P, or

- we land up on a vertex in a cycle of C.

Note that not both can happen, as one must happen first. Moreover, indeed one will happen, as the total number of edges in P and C is n, and if P and C have respectively k and $n-k$ edges, then, providing P never repeats a vertex, P and C have between them $k+1+n-k = n+1$ vertices on their edges, implying that some vertex must be on both P and C. Thus we define $\alpha(P,C)$ as (P',C'), where

- in the first case, the cycle determined by the return to P is moved from P to C (see Figure 3.9 – P is shown as a solid line, C as a union of dashed heavy cycles), and

- in the second case, the cycle of C that first touches P is removed from C and inserted appropriately into P (see Figure 3.10 – again, P is shown as a solid line, C as a union of dashed heavy cycles).

FIGURE 3.9: Cutting a cycle from P

FIGURE 3.10: Moving a cycle into P

It is straightforward (and easy!) to check that α^2 is the identity map on $\mathcal{F}_{i,j}$ and that for any $(P,C) \in \mathcal{F}_{i,j}$, $\mathrm{wt}(\alpha(P,C)) = -\mathrm{wt}(P,C)$. Thus $\mathrm{wt}(\mathcal{F}_{i,j}) = 0$, concluding our proof of the Cayley–Hamilton theorem. ∎

3.2 Graphs and Other Branches of Mathematics

While we have seen examples of how to use graphs and directed graphs as models in mathematics, it should not be surprising that other discrete structures and other areas of mathematics can be used very successfully to model graphs and directed graphs. As we haven't examined other discrete structures in much detail yet, we'll devote this section to connections with other areas of mathematics.

3.2.1 Graphs and Topology

We know that once we represent a graph in a Euclidean space, we have associated a topological structure to a graph, and we can then apply any results as we see fit from the deep theory of topology. Here is a simple result, one that you could prove otherwise. We consider it as an example of how simple results in one area can yield results in another.

Consider the following lemma from chromatic theory, and a proof via some elementary topology:

Lemma 3.6 *Let $G = (V, E)$ be a finite simple graph. Then for any $0 \leq k \leq \chi(G)$, there is a subgraph G_k of G that is k–chromatic.*

Proof: We begin by modeling a graph with a graph! We form a new graph $\mathrm{Subgr}(G)$ whose vertices are the spanning subgraphs of G, with an edge between $H_1 = (V, E_1)$ and $H_2 = (V, E_2)$ if and only if $|E_1 \triangle E_2| = 1$, i.e. the two subgraphs differ in just one edge. You can verify that $\mathrm{Subgr}(G)$ is connected. We then take any representation $(\rho, \{\phi_e : e \in E\})$ (such as described earlier) of G in $\mathbb{R}^3$ such that edges meet only at vertices (here we identify the vertices and edges with their images under the representation). Let $\mathbf{G}$ denote the image of G in $\mathbb{R}^3$ under the representation. Finally, we define a map $f : \mathbf{G} \to [0, \chi(G)]$ such that for any $H \in \mathrm{Subgr}(G)$,

- $f(\phi(H)) = \chi(H)$, and

- for any $\mathbf{x} = \phi_e(t)$ ($e = \{H_1, H_2\}$, $t \in (0, 1)$), we extend f *linearly*, i.e. $f(\mathbf{x}) = (1 - t) \cdot f(\phi_e(H_1)) + t \cdot f(\phi_e(H_2))$.

Now it is easy to see that f is continuous on the compact set $\mathbf{G}$. It is also not hard to see that for any edge $e = \{H_1, H_2\}$ of a simple graph G, $|f(\rho(H_1)) - f(\rho(H_1))| = 0$ or 1, as the deletion of an edge of a graph either leaves the chromatic number the same or decreases it by exactly one. It follows if $f(\mathbf{x})$ is an integer l, then $f(\mathbf{x}) = f(\rho(H))$ for some spanning subgraph H of G.

Now $\mathrm{Subgr}(G)$ is connected, and this implies that $\mathbf{G}$ is (path) connected in $\mathbb{R}^3$. The result now follows as the image of a connected set $\mathbf{G}$ under a continuous map is connected, so for any integer $k \in [0, \chi(G)]$, there must be an $\mathbf{x} \in \mathbf{G}$ such that $f(\mathbf{x}) = k$, and as noted before, this implies that for some $H \in \mathrm{Subgr}(G)$ we have $\chi(H) = f(\rho(H)) = k$. ∎

While this example might seem like going after a mosquito with a cannon, the principle of using topological connectivity to show the existence of a subgraph with certain properties can be a useful one; see [63] for an example that relates the topological property of being simply connected to spanning trees of a graph.

3.2.2   Graphs and Algebra

We now turn to see how algebra can serve to model graphs in useful ways. We begin with an example from matrix theory. A **vertex cover** of a simple graph $G = (V, E)$ is a collection of vertices $B \subseteq V$ such that every edge of G has at least one end in B; we let $\tau(G)$ denote the **vertex cover number**, the minimum cardinality of a vertex cover of G. G is τ–**critical** if $\tau(H) < \tau(G)$ for every proper subgraph H of G.

Example: The graph C_{2n+1} is τ–critical for all $n \geq 1$ as $\tau(C_{2n+1}) = n + 1$ while $\tau(C_{2n+1} - e) = \tau(P_{2n+1}) = n$.   △

It is not obvious whether there are infinitely many τ–critical graphs with a given τ value. Erdös showed that, in fact, there are only finitely many such graphs; the proof we provide is due to Lovász [65].

Theorem 3.7 *If a simple graph $G = (V, E)$ is τ–critical with $\tau(G) = t$, then G has size at most $\binom{t+1}{2}$. Hence there are only finitely many such τ–critical graphs.*

Proof: [65] Let $t = \tau(G)$. Choose for each vertex v of G a (column) vector $\mathbf{v} \in \mathbb{R}^t$ such that $\{\mathbf{v} : v \in G\}$ is in *general position*, that is, any t of them are linearly independent (see Exercise 3.26). For $e = \{x, y\} \in E$, form the $t \times t$ matrix

$$A_e = \mathbf{x}\mathbf{y}^T + \mathbf{y}\mathbf{x}^T.$$

Note that A_e is symmetric, so it sits in the vector space Symm_t of all symmetric $t \times t$ matrices, which has dimension $\binom{t+1}{2}$ (see Exercise 3.27).

We now claim that $\{A_e : e \in E\}$ is linearly independent (in Symm_t). For suppose not; then for some $f = \{u, v\} \in E$, we have

$$A_f = \sum_{e \in E - \{f\}} \lambda_e A_e$$

for some $\lambda_e \in \mathbb{R}$. As G is τ–critical with $\tau(G) = t$, there is a set B of $t - 1$ vertices of G that cover all edges of $G - f$. Let $\mathbf{c} \in \mathbb{R}^t$ be a nonzero vector orthogonal to the set $\{\mathbf{b} : b \in B\}$. Then for any edge $e = \{x, y\} \neq f$, at least one of x or y belongs to B, so that $\mathbf{c}^T \mathbf{x} = 0$ or $\mathbf{c}^T \mathbf{y} = 0$. Thus for $e \in G - f$,

$$\begin{aligned}
\mathbf{c}^T A_e \mathbf{c} &= \mathbf{c}^T \mathbf{x} \mathbf{y}^T \mathbf{c} + \mathbf{c}^T \mathbf{y} \mathbf{x}^T \mathbf{c} \\
&= 2(\mathbf{c}^T \mathbf{x})(\mathbf{c}^T \mathbf{y}) \\
&= 0.
\end{aligned}$$

It follows that

$$\mathbf{c}^T A_f \mathbf{c} = \mathbf{c}^T \left(\sum_{e \in E - \{f\}} \lambda_e A_e \right) \mathbf{c} = 0.$$

On the other hand,

$$\mathbf{c}^T A_f \mathbf{c} = 2(\mathbf{c}^T \mathbf{u})(\mathbf{c}^T \mathbf{v}) \neq 0$$

as (i) $\{\mathbf{v} : v \in G\}$ is in general position, and (ii) $\{\mathbf{b} : b \in B\}$ is a basis for the orthogonal complement of $\text{Span}(\mathbf{c})$ implies that neither $\mathbf{u}$ nor $\mathbf{v}$ is orthogonal to $\mathbf{c}$. This contradiction implies that indeed the set $\{A_e : e \in E\}$ is linearly independent in Symm_t, so it can contain at most $\dim(\text{Symm}_t) = \binom{t+1}{2}$ many vectors. It follows that $|E| \leq \binom{t+1}{2}$. That there are therefore finitely many such τ–critical graphs follows from the fact that no τ–critical graph G can have an isolated vertex. ∎

We now turn to a problem, similar to colourings, that illustrates the use of linear systems in graph theory. For a simple graph $G = (V, E)$, let $cl(G)$ denote the **clique partition number**, that is, the minimum number of cliques whose edge sets partition the edges of G. Clearly $cl(G) \leq |E|$ as one can take each edge of G as a clique to partition the edges of G. Sometimes, however, you can do much better; in the example of Figure 3.11, F can be partitioned

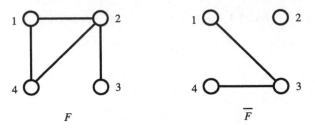

FIGURE 3.11: A graph F and its complement

into 2 cliques, as can $\overline{F}$, the complement of F (given a simple graph G, the **complement** of G, $\overline{G}$, is the graph on vertex set $V = V(G)$ such that for all distinct u, $v \in V$, uv is an edge of $\overline{G}$ if and only if uv is <u>not</u> an edge of G). Note that in this example, $cl(F) + cl(\overline{F}) = 2 + 2 = 4 \geq |V(F)|$.

Similarly, $cl(K_{n,m}) + cl(\overline{K_{n,m}}) = nm + 2 > n + m = |V(K_{n,m})|$. On the other hand, we have $cl(K_n) + cl(\overline{K_n}) = 1 < |V(K_n)|$ if $n > 1$. Erdös and de Bruijn [34] proved via graph theory that indeed $G = K_n$ or $\overline{K_n}$ are the only "bad" cases; we prove here the result with a new proof, via linear systems.

Theorem 3.8 *For any simple graph $G = (V, E)$, $cl(G) + cl(\overline{G}) \geq |V|$ provided neither G nor $\overline{G}$ are complete.*

Proof: Let $\theta_1, \ldots, \theta_r$ be the vertex sets of a partition of the edges of G into cliques, and let $\gamma_1, \ldots, \gamma_l$ be the vertex sets of a partition of the edges of $\overline{G}$ into cliques. We want to show that $r + l \geq n = |V|$.

For each vertex $v_i \in V$, introduce a variable x_i and set

$$\widehat{\theta_i} = \sum_{v_i \in \theta_i} x_i$$

and

$$\widehat{\gamma_i} = \sum_{v_j \in \gamma_j} x_j.$$

Now

$$(\widehat{\theta_i})^2 = \sum_{v_j \in \theta_i} x_j^2 + 2 \sum_{\{v_j v_k\} \in \theta_i} x_j x_k.$$

A similar formula holds for $(\widehat{\gamma}_i)^2$. It follows that

$$\sum_{i=1}^{r}(\widehat{\theta}_i)^2 + \sum_{j=1}^{l}(\widehat{\gamma}_j)^2 = \sum_{i=1}^{n}\rho_i x_i^2 + 2\sum_{i\neq j}x_i x_j, \qquad (3.4)$$

where ρ_i is the number of cliques among $\theta_1,\ldots,\theta_r,\gamma_1,\ldots,\gamma_l$ that contain v_i.

Note that if $\deg_G(v_i) = n-1$ then as G is not complete, v_i is an endpoint in at least two θ_js, so $\rho_i \geq 2$. If $\deg_G(v_i) = 0$, then $\deg_{\overline{G}}(v_i) = n-1$ and again $\rho_i \geq 2$. Finally, if $\deg_G(v_i) \neq 0$, $n-1$ then v_i is an endpoint of an edge in both G and $\overline{G}$, so that $\rho_i \geq 2$ here as well. Thus in all cases, $\rho_i \geq 2$. Thus from (3.4) we have

$$\sum_{i=1}^{r}(\widehat{\theta}_i)^2 + \sum_{i=1}^{l}(\widehat{\gamma}_i)^2 = \left(\sum_{i=1}^{n}x_i\right)^2 + \sum_{i=1}^{n}(\rho_i - 1)x_i^2. \qquad (3.5)$$

Now consider the linear system

$$\widehat{\theta}_1 = 0$$
$$\widehat{\theta}_2 = 0$$
$$\cdots$$
$$\widehat{\theta}_r = 0$$
$$\widehat{\gamma}_1 = 0$$
$$\widehat{\gamma}_2 = 0$$
$$\cdots$$
$$\widehat{\gamma}_l = 0.$$

This is a homogeneous linear system with $r+l$ equations and n unknowns.

If $r+l < n$, then the homogeneous linear system has fewer equations than unknowns, so it has a nontrivial solution $(x_1^*,\ldots,x_n^*) \in \mathbb{R}^n$. But then from (3.5)

$$0 = \left(\sum_{i=1}^{n}x_i^*\right)^2 + \sum_{i=1}^{n}(\rho_i - 1)(x_i^*)^2.$$

The first term on the right hand side is clearly nonnegative, and the second is positive as each $\rho_i - 1$ is at least 1 and some x_i^* is nonzero. Thus the right side is positive, while the left side is 0, a contradiction. Thus we conclude

$r + l \geq n$. By choosing a minimal clique partition for both G and $\overline{G}$, we derive that $cl(G) + cl(\overline{G}) \geq |V|$. ∎

We have seen how reliability is essentially a polynomial model of (multi)graph that encodes "connectedness" under the probabilistic model suggested. The fact that the model is a polynomial suggests a number of different approaches. For example, polynomials over any field form a vector space, and vectors can be expanded in terms of *any* basis. Thus we could consider different bases for the polynomial space $\mathbb{R}[p]$ and expand the reliability polynomial in terms of these bases. A variety of different expansions have been found to be useful, particularly in the problem of estimating the reliability of a graph quickly (the problem of calculating the reliability polynomial is known to be intractable [31]). The set of polynomials $\{p^{m-i}(1-p)^i : i = 0, 1, \ldots, m\}$ form a basis for the subspace of $\mathbb{R}[p]$ of polynomials of degree at most m, and it is this basis in which we originally expanded $\text{Rel}(G, p)$. Of course, there is a standard expansion in terms of powers of p. Here are some of the expansions of the reliability polynomial:

$$\text{Rel}(G, p) = \sum_{i=n-1}^{m} (-1)^{i-n+1} S_i p^i \quad \text{(S–form)}$$

$$= \sum_{i=0}^{m-n+1} F_i p^{m-i}(1-p)^i \quad \text{(F–form)}$$

$$= \sum_{i=n-1}^{m} N_i p^i (1-p)^{m-i} \quad \text{(N–form)}$$

$$= p^{n-1} \sum_{i=0}^{m-n+1} H_i (1-p)^i \quad \text{(H–form)}.$$

Example: With G being the graph in Figure 3.12, the S–, F–, N– and H–forms

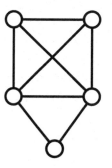

FIGURE 3.12: A graph G

are, respectively,

$$
\begin{aligned}
\mathrm{Rel}(G,p) &= 40p^4 - 109p^5 + 115p^6 - 55p^7 + 10p^8 \\
&= 1(1-p)^0 p^8 + 8(1-p)^1 p^7 + 28(1-p)^2 p^6 + 51(1-p)^3 p^5 + \\
&\quad 40(1-p)^4 p^4 \\
&= 40p^4(1-p)^4 + 51p^5(1-p)^3 + 28p^6(1-p)^2 + 8p^7(1-p)^1 + \\
&\quad 1p^8(1-p)^0 \\
&= p^4(1(1-p)^0 + 4(1-p)^1 + 10(1-p)^2 + 15(1-p)^3 + \\
&\quad 10(1-p)^4)
\end{aligned}
$$

$\triangle$

As the reliability polynomials lie in $\mathbb{Q}[p]$, it is not hard to see that each of the coefficients in each form is rational. It turns out that all the sequences $\langle S_i \rangle$, $\langle F_i \rangle$, $\langle N_i \rangle$ and $\langle H_i \rangle$ consist of nonnegative integers. In fact, these sequences have combinatorial significance. You should be able to verify that N_i is the number of subsets of edges of size i that form a spanning connected subgraph of G. That the H_is are nonnegative integers will be shown later, when we learn more about simplicial complexes and their models.

3.2.3 Graphs and Analysis

The connection to polynomials suggests that there might be interest in studying the roots of reliability polynomials. Such an investigation has yielded interesting conjectures to explore. An examination of the S–, F–, N– and H–sequences for small graphs led to a number of conjectures [18, 31].

Conjecture 3.9 ([18]) *For any connected graph G, the S–, F– and N–sequences are log concave, and hence unimodal.*

Conjecture 3.10 ([31]) *For any connected graph G, the H–sequence is log concave, and hence unimodal.*

Example: For $\text{Rel}(K_4, p)$, the S–, F–, N– and H–sequences are, respectively, $(16, 33, 24, 6)$, $(1, 6, 15, 16)$, $(16, 15, 6, 1)$ and $(1, 3, 6, 6)$, all of which are log concave, and hence unimodal. As well, the previous example shows that for the graph G of Figure 3.12, the S–, F–, N– and H–sequences are, respectively, $(40, 109, 115, 55, 10)$, $(1, 8, 28, 51, 40)$, $(40, 51, 28, 8, 1)$ and $(1, 4, 10, 15, 10)$, all of which are log concave, and hence unimodal.   △

These conjectures, if true, also have practical applications in combinatorial reliability theory, as they would imply stronger constraints on the coefficients than previously known (more about this in the section on simplicial complexes). The difficulty in proving the unimodality of such sequences is that, unlike many other combinatorial sequences, the sequences are not in general symmetric, and, moreover, the location of "peak" is not known (and not, in general, in the middle).

Given a graph G and edge e of G, a **subdivision of edge e** is a replacement of e by a path of some positive length. A **subdivision of graph G** is a graph formed from G by a sequence of subdivisions of edges. We will prove that for any graph G, there is a subdivision G' of G such that $\text{Rel}(G', p)$ has all real roots. What does a polynomial having all real roots imply about the coefficients of a polynomial? It actually implies a lot, via an inequality due to Newton (c.f. [32, pg. 270–271] and [51, pg. 104–105]). The result is useful well beyond reliability, and the beautiful proof will use nothing more than first year calculus!

Theorem 3.11 (Newton's Theorem) *If a polynomial*

$$b_0 + b_1 x + \cdots + b_k x^k$$

of degree $k \geq 1$ with positive coefficients has only real roots, then

$$b_i^2 \geq \frac{i+1}{i} \frac{k-i+1}{k-i} b_{i-1} b_{i+1} \qquad (1 \leq i \leq k-1)$$

and hence the sequence $\langle |b_0|, |b_1|, \ldots, |b_k| \rangle$ is strictly log concave and thus unimodal.

Proof: First we note that if a polynomial $f = f(x)$ of degree $d \geq 1$ with real coefficients has all real roots, then so does its derivative, f'. Why is this true? If f has all real roots, then for some positive integer l, positive integers $k_1, k_2, \ldots k_l$ and real numbers $\alpha_1 < \alpha_2 < \cdots < \alpha_l$, we have

$$f = (x - \alpha_1)^{k_1} (x - \alpha_2)^{k_2} \cdots (x - \alpha_l)^{k_l},$$

where $\sum k_i = d$. It is not hard to check, via the product rule for derivatives, that f' has a root of multiplicity (as least) $k_i - 1$ at each α_i. Moreover, in each of the $l - 1$ intervals $(\alpha_1, \alpha_2), \ldots, (\alpha_{l-1}, \alpha_l)$ there is a root of f' as well, by Rolle's theorem (that is, between the roots of f, f has a local minimum or maximum). It follows that we have found $(k_1 - 1) + \cdots + (k_l - 1) + l - 1 = d - 1$ many real roots of f'. However, f' has degree $d - 1$, and hence we have found all of its roots, and they are real!

So now suppose that a polynomial

$$g(x) = b_0 + b_1 x + \cdots + b_k x^k$$

of degree k with positive coefficients has all real roots. Fix $i \in \{1, \ldots, k-1\}$. Then from the previous observation so does g', and by the same argument, g'', g''' and so on – *all* derivatives $g^{(i)}$ of g have all real roots (by convention, $g^{(0)} = g$). Thus

$$h(x) = g^{(i-1)}(x) = b_{i-1}(i-1)! + b_i \frac{i!}{1!} x + \cdots + b_k \frac{k!}{(k-(i-1))!} x^{k-(i-1)}$$

has all real roots. It is not hard to check that the polynomial

$$H(x) = x^{k-i+1} h(1/x) = b_k \frac{k!}{(k-i+1)!} + \cdots + b_i \frac{i!}{1!} x^{k-i} + b_{i-1}(i-1)! x^{k-i+1},$$

which reverses the order of the coefficients of h, also has all real roots (see Exercise 3.31).

Again, by the same reasoning, all of the derivatives of H have all real roots. The key now is to differentiate down until we reach a quadratic – where we *know* how to check for realness of the roots. Thus we see that

$$H^{k-i-1}(x) = b_{i+1}\frac{(i+1)!(k-i-1)!}{2} + b_i i!(k-i)!x + b_{i-1}\frac{(i-1)!(k-i+1)!}{2}x^2$$

has all real roots. But this quadratic has all real roots just in case its discriminant

$$(b_i i!(k-i)!)^2 - 4b_{i-1}\frac{(i-1)!(k-i+1)!}{2}b_{i+1}\frac{(i+1)!(k-i-1)!}{2}$$

is nonnegative, so that

$$(b_i i!(k-i)!)^2 \geq b_{i-1}b_{i+1}(i-1)!(i+1)!(k-i-1)!(k-i+1)!.$$

A little bit of cancelling shows that this is equivalent to

$$b_i^2 \geq \frac{i+1}{i}\frac{k-i+1}{k-i}b_{i-1}b_{i+1}$$

which implies that

$$b_i^2 > b_{i-1}b_{i+1},$$

that is, the sequence is (strictly) log concave, and hence unimodal. ∎

We point out that if the coefficients of $g(x)$ alternate in sign (and are nonzero) then from the argument above they are strictly log concave and are therefore unimodal in absolute value. We can now return to our main theorem on unimodality for reliability polynomials (for a more general argument in the setting of matroids, see [19]).

Theorem 3.12 *For any connected graph G, there is a subdivision G' of G such that log concavity holds for all the sequences associated with $Rel(G', p)$.*

Proof: Let G have order n and size m (we assume that G is loopless – loops can be safely removed without changing the reliability).

It will be easier for us to get rid of 0 as a root by considering the polynomial

$$f_G(p) = \frac{\mathrm{Rel}(G,p)}{p^{|V(G)|-1}},$$

since we have already seen that the lowest power of p occurring with nonzero coefficient in $\mathrm{Rel}(G,p)$ is p^{n-1}, that is, 0 is a root of multiplicity $n-1$ in $\mathrm{Rel}(G,p)$ (the coefficient of p^{n-1} in $\mathrm{Rel}(G,p)$ is the number of spanning trees of G, and hence positive). We shall show that G has a subdivision G' such that the roots of $f_G(p)$ are all real and distinct; this will show that $\mathrm{Rel}(G,p)$ has all real roots, and hence, via Theorem 3.11, the coefficients of $\mathrm{Rel}(G,p)$ in standard form are unimodal in absolute value. The fact that the other coefficients will then have all real roots is left as an exercise (see Exercise 3.35).

We induct on the number of edges in G. If G has $n-1$ edges (this is the base case, as G is connected) then G is a tree with reliability p^{n-1}, so $f_G(p) = 1$ and the result is trivial. Thus we assume that $m \geq n$ and that the result holds for graphs with fewer number of edges. Let e be an edge of G such that $G-e$ is connected (any edge in a cycle of G will do). We can assume inductively that we have a subdivision $(G-e)'$ such that $f_{(G-e)'}$ has all roots real and distinct. Let $(G-e)'$ have order N and size M. Then we can list the roots of $f_{(G-e)'}$ as $\alpha_1 < \alpha_2 < \cdots < \alpha_{M-N+1}$, and we know that all of these are positive (as the coefficients of $\mathrm{Rel}((G-e)',p)$ and hence $f_{(G-e)'}(p)$ alternate in sign). Thus we can **interlace** these roots by choosing β_i, $i = 0, \ldots, M-N+1$ such that $0 < \beta_0 < \alpha_1 < \beta_1 < \alpha_2 < \ldots < \alpha_{M-N} < \beta_{M-N} < \alpha_{M-N+1} < \beta_{M-N+1}$ so that

$$\mathrm{sign}(f_{(G-e)'}(\beta_i)) = (-1)^i.$$

(Note that $f_{(G-e)'}(0) > 0$ implies that $f_{(G-e)'}(x) > 0$ for $x < \alpha_1$.)

We now form graph $\widehat{G_l}$ from $(G-e)'$ by adding back in edge e and subdividing it into a path of length l. It is not hard to verify by induction that

$$\mathrm{Rel}(G_l,p) = lp^{l-1}(1-p)\mathrm{Rel}((G-e)',p) + p^l\mathrm{Rel}(F,p),$$

where F is the graph formed from $(G-e)'$ by adding in edge e and then contracting it. By dividing through by the appropriate power of p, we derive

$$f_{G_l}(p) \quad = \quad l(1-p)f_{(G-e)'}(p) + f_F(p). \tag{3.6}$$

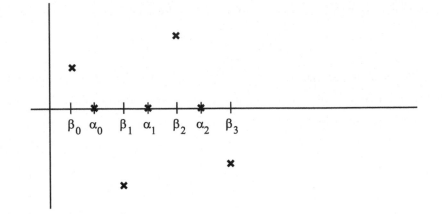

FIGURE 3.13: The sign of $f_{(G-e)'}$ (indicated by an x)

Note that the degrees of the polynomials $f_{G_l}(p)$, $f_{(G-e)'}(p)$ and $f_F(p)$ are respectively $M - N + 2$, $M - N + 1$ and $M - N + 2$.

Now by choosing l large enough, we can ensure that the absolute values of $l(1-p)f_{(G-e)'}(\beta_i)$ dominate $f_F(\beta_i)$. It follows that

$$\text{sign}(f_{G_l}(\beta_i)) = \text{sign}(l(1-p)f_{(G-e)'}(\beta_i))) = -\text{sign}(f_{(G-e)'}(\beta_i))) = (-1)^{i+1}$$

as the β_i are positive, so $f_{G_l}(p)$ will alternate in sign at the β_is. As $f_{G_l}(p)$ is continuous, by the Intermediate Value Theorem, it will have a root in each of the intervals $(\beta_0, \beta_1), (\beta_1, \beta_2), \ldots, (\beta_{M-N}, \beta_{M-N+1})$.

Now as $\text{sign}(f_{G_l}(\beta_{M-N+1})) = (-1)^{M-N+1}$ and $f_{G_l}(p)$ has leading term with coefficient of sign $(-1)^{M-N+2}$, $f_{(G_l}(p)$ has another root in (β_{M-N+1}, ∞). Thus we have found $M - N + 1 + 1 = M - N + 2$ distinct roots for $f_{G_l}(p)$, a polynomial of degree $M - N + 2$. It follows that $fG_l(p)$ has distinct real roots, and we are done. ∎

Another conjecture from [18] is the following:

Conjecture 3.13 ([18]) *For any connected graph G, the roots of $\text{Rel}(G, p)$ lies in $|p - 1| \leq 1$.*

Various evidence was presented for the conjecture in [18], such as showing that every graph has a subdivision whose roots lie in this disk, and any

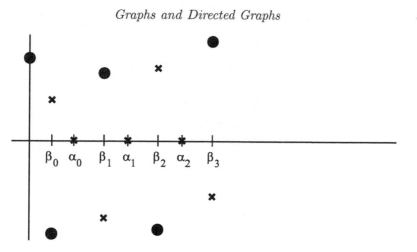

FIGURE 3.14: The sign of f_{G_l} (indicated by a filled circle)

real root of a reliability polynomial of a connected graph lies in $[0, 2]$. Connected **series–parallel graphs** are formed recursively, starting from a tree, by replacing a single edge $e = xy$ by two edges xy and zy in series (where z is a new vertex) or by two parallel edges, with the endpoints x and y (we say "series–parallel graph" but we really mean "series–parallel multigraph" as there can be multiple edges). An example of a series–parallel graph is shown in Figure 3.2.3.

Wagner [102] proved Conjecture 3.13 for series-parallel graphs in a lengthy (24 page), involved paper. But a theoretical physicist, Alan Sokal, reproved this result using an ingenious concept – sometimes it is easier to prove a seemingly more difficult extension of a conjecture than the simpler looking original version! To wit, Sokal extended Conjecture 3.13 to the more general form of reliability:

Conjecture 3.14 ([81]) *For any connected graph G, if $|1 - p_e| > 1$ for all edges e of G, then $Rel(G, \{p_e : e \in E\}) \neq 0$.*

This statement is seemingly more complex as the polynomial has many variables, rather than just one. But Sokal goes on to reprove Wagner's theorem in this stronger, multivariate setting. First it is not hard to see (Exercise 3.36) that if G' is the graph formed from G by a parallel operation, replacing edge

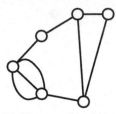

FIGURE 3.15: A series–parallel graph

$e = xy$ by the edges $e_1 = xy$ and $e_2 = xy$, and G'' is the graph formed from G by a series operation, replacing edge $e = xy$ by the edges $e_1 = xz$ and $e_2 = zy$ (z a new vertex), then

$$\text{Rel}(G', p_e, p_1, p_2) \quad = \quad \text{Rel}(G, p_e \leftarrow 1 - (1 - p_1)(1 - p_2))$$

and

$$\text{Rel}(G'', p_e, p_1, p_2) \quad = \quad (1 - (1 - p_1)(1 - p_2))\text{Rel}(G, p_e \leftarrow \tfrac{p_1 p_2}{p_1 + p_2 - p_1 p_2})$$

where edges e_1 and e_2 have edge probabilities p_1 and p_2, respectively, and the arrow indicates a substitution of the expression on the right for the variable on the left.

Now Sokal notes that if $|1 - p_i| > 1$ for $i = 1$ and 2, then the same is true for $1 - (1 - p_1)(1 - p_2)$ and $\tfrac{p_1 p_2}{p_1 + p_2 - p_1 p_2}$. The former is easy, as

$$|1 - (1 - (1 - p_1)(1 - p_2))| = |(1 - p_1||(1 - p_2| > 1 \cdot 1 = 1.$$

For the latter, see Exercise 3.37. As Conjecture 3.14 clearly holds for trees (as the reliability of a tree is the product of the edge probabilities), by induction, we see that the Conjecture 3.14 holds for all series-parallel graphs, and hence Conjecture 3.13 is true for series-parallel graphs. The more complex, multivariate conjecture is the way to go here! Why? Having different variables around for edges allows us to do induction of the variables, which we can't do if all the variables are set equal to begin with. The lesson is not to shy away from broader generalizations - generalizing may indeed be more than interesting, it may be very, very helpful!

Just a note – Sokal and Gordon Royle showed [83] that Conjecture 3.13 is false in general, as they found graphs whose roots lie outside the disk $|p-1| \leq$

1, but only by the slimmest of margins (the maximum of $|1-p|$ for a root was at about 1.04).

3.2.4 Graphs and Logic

We have already discussed the problem of determining the chromatic number of $\mathbb{G}^2$, the graph on $\mathbb{R}^2$ where edges correspond to vertices at distance 1. It seems clear that any k–colouring of $\mathbb{G}^2$ must involve colouring the entire plane. On the other hand, how about proving that $\mathbb{G}^2$ can't be coloured with k colours? Need we consider the entire infinite (indeed, uncountable!) graph before we can provide a proof that it can't be coloured with a given number of colours? Thankfully, the answer is no – if $\mathbb{G}^2$ can't be coloured with k colours, then there is a *finite* induced subgraph that can't be coloured with k colours as well, and hence there is a proof that the graph $\mathbb{G}^2$ requires more colours. This fact is due to a more general result (due to Erdös and de Bruijn [34]) on traveling from the finite to the infinite via first-order logic. Erdös–de Bruijn Theorem

Theorem 3.15 ([34]) *Let $G = (V, E)$ be an infinite simple graph. Then G can be k–coloured if and only if every finite (induced) subgraph of G can be coloured with k colours.*

Proof: To prove this result, we need to model G in first-order logic. One direction is obvious – if G has a k–colouring $\pi : V \to \{1, \ldots, k\}$, then so does every subgraph H, a k-colouring being the restriction of π to the vertices of H.

To prove the converse, we use an extension of the first-order theory of graphs. We take a binary predicate R for the edge relation, take a constant c_v for every vertex v of G (thus, there are infinitely many such constants), and introduce k unary predicates $C_1, \ldots, C_k$. We now write down a set of sentences $\mathcal{T}$ that "encodes" that G has a k–colouring. The sentences of $\mathcal{T}$ are:

i. $(\forall x)(\neg R(x, x))$,

ii. $(\forall x)(\forall y)(R(x, y) \to R(y, x))$,

iii. $R(c_v, c_u)$ for all edges $\{v, u\} \in E$,

iv. $\neg R(c_v, c_u)$ for all $\{v, u\} \notin E$,

v. $(\forall x)(C_1(x) \vee C_2(x) \vee \ldots \vee C_k(x))$,

vi. $(\forall x)(\neg(C_1(x) \wedge C_2(x)) \wedge \neg(C_1(x) \wedge C_3(x)) \wedge \ldots \wedge \neg(C_{k-1}(x) \wedge C_k(x)))$, and

vii. $(\forall x)(\forall y)(R(x, y) \rightarrow (\neg(C_1(x) \wedge C_1(y)) \wedge \neg(C_2(x) \wedge C_2(y)) \wedge \ldots \wedge \neg(C_k(x) \wedge C_k(y))))$.

(In essence, (i) states that there are no loops, (ii) that the edge relation R is symmetric, (iii) and (iv) encode the particular graph G, and (v)-(vii) ensure that C_1 through C_k are colour classes.)

If we have a model M of $\mathcal{T}$, then the vertices corresponding to the interpretations of the constants will form a subgraph $H = (V', E')$ isomorphic to G. Moreover, for $v \in V$, let $\hat{v}$ denote the unique $i \in \{1, \ldots, k\}$ such that $C_i(c_v)$ is true in M. Then $\pi : V' \rightarrow \{1, \ldots, k\} : v \mapsto \hat{v}$ yields a k–colouring of G. So all that remains to show is that $\mathcal{T}$ has a model.

Recall the Satisfiability Theorem from first-order logic: a consistent set of sentences (i.e. one from which you can prove $\phi \vee (\neg\phi)$ for some (or any!) sentence ϕ) has a model. Therefore we need only show that $\mathcal{T}$ is consistent, which is equivalent to showing that every finite subset of $\mathcal{T}$ is consistent. It suffices, by the Compactness Principle, to show that any finite subset of $\mathcal{T}$ has a model (the existence of a model ensures consistency).

Let S be any finite subset of $\mathcal{T}$; we shall show that S has a model. Let S denote the collection of $v \in V$ such that some sentence of form (iii) or (iv) contains the constant c_v; S is finite as $\mathcal{S}$ is. Extend S to S' by ensuring that the sentences (i), (ii) and (v)-(vii) are present, as well as all sentences of type (iii) and (iv) that contain a pair of constants c_s, $c_{s'}$ where $s, s' \in S$. The set of sentences S' is still finite, and if we take a k–colouring of the subgraph of G induced by S, we find we have a model of S' with the obvious interpretations. Thus every finite subset of $\mathcal{T}$ has a model, so we conclude that $\mathcal{T}$ has a model, that is, G is k–colourable. ∎

The upshot of this theorem is that if the graph $\mathbb{G}^2$ cannot be coloured by say 5 colours, then there will exist a finite set of points such that the subgraph of G induced by these points is not 5–colourable.

3.2.5 Graphs and Probability

It may seem strange that there are cases when it is unknown how to construct a discrete structure with a given property, and yet by viewing graphs under a probabilistic model, one can show that indeed such a discrete structure need exist, even though you can't actually exhibit one! We'll demonstrate the technique with one of the best known (and first) uses of the probabilistic approach in graph theory.

One way to have a large chromatic number is to have large cliques within a graph, as clearly the chromatic number is at least the clique number. Are there any other ways to increase the chromatic number other than increasing the clique number? Moreover, can we increase the chromatic number without having any "short" cycles? Of course, we need some cycles in the graph, as a cycle–free graph is 2–colourable (indeed for the same reason we need some odd cycles in the graph). Try to construct a 4–colourable graph without any cycles of lengths at most 6, and you will see how difficult the problem is.

Erdös [36], in one of the earliest uses of the probabilistic argument in graph theory, proved the existence of graphs with high chromatic number without any small cycles. The argument we provide is similar to that found in [14]. Note that this argument, like many probabilistic arguments in graph theory, could be restated as a counting problem, but this goes against the grain as we are seeking to highlight the connections between combinatorics and other disciplines.

We shall use some standard inequalities, most derived from **Stirling's Formula**:

$$n! = \left(\frac{n}{e}\right)^n \sqrt{2\pi n}e^{\alpha/12n}, \text{ where } \alpha \in [0,1] \text{ depends on } n.$$

The important inequalities (with $x \le b \le a$ and $y \le a$):

$$\binom{a}{b} \le \left(\frac{ea}{b}\right)^b, \tag{3.7}$$

$$\frac{\binom{a-x}{b-x}}{\binom{a}{b}} \le \left(\frac{b}{a}\right)^y, \text{ and} \tag{3.8}$$

$$\frac{\binom{a-y}{b}}{\binom{a}{b}} \le e^{-(b/a)y}. \tag{3.9}$$

Theorem 3.16 *Let k, $g \ge 4$. Then there is a graph G that has no cycles of length less than g that is not k–colourable.*

Proof: Our sample space is the collection Ω of all graphs of order n on a fixed set, with $m = 2k^3n$ edges (such graphs are often called **sparse** as they have relatively few edges). Each graph in our sample space is equally likely to be chosen (with probability $\left(\binom{\binom{n}{2}}{m}\right)^{-1}$.) We shall estimate first the expected number of cycles of length $l < g$ (in all inequalities, we assume that n is sufficiently large, and that n is a multiple of k). Let L_l denote the event that a graph G in the space has a cycle of length l. Then

$$
\begin{aligned}
E(L_l) &= \frac{n(n-1)\cdots(n-l+1)}{2l} \frac{\binom{\binom{n}{2}-l}{m-l}}{\binom{\binom{n}{2}}{m}} \\[2mm]
&< \frac{n^l}{2l}\left(\frac{m}{\binom{n}{2}}\right)^l \\[2mm]
&\le \frac{n^l}{2l}\left(\frac{4k^3}{n-1}\right)^l \\[2mm]
&\le \frac{n^l}{2l}\left(\frac{5k^3}{n}\right)^l \\[2mm]
&\le \frac{(5k^3)^l}{2l}
\end{aligned}
$$

It follows that

$$
\begin{aligned}
\sum_{l=3}^{g-1} E(L_l) &< \sum_{l=3}^{g-1} \frac{(5k^3)^l}{2l} \\[2mm]
&\le \frac{(5k^3)^g}{6}
\end{aligned}
$$

so that if E_1 is the event that there are at most $\frac{(5k^3)^g}{3}$ cycles of length less than g in G, then

$$\text{Prob}(E_1) \geq \frac{1}{2}, \tag{3.10}$$

as otherwise the probability that G has more than $\frac{(5k^3)^g}{3}$ cycles of length less than g is at least $1/2$, and this adds more than $\frac{(5k^3)^g}{6}$ to the sum above, a contradiction.

Now let's consider the event I_s that some set of $q = n/k$ vertices have s edges, for $s \leq S = \dfrac{(5k^3)^g}{3}$. We see that

$$
\begin{aligned}
E(I_s) &= \binom{n}{q}\binom{\binom{q}{2}}{s}\binom{\binom{n}{2} - \binom{q}{2}}{m - s} \\
&\leq \left(\frac{en}{q}\right)^q \left(\frac{eq^2}{2}\right)^s e^{-m\binom{q}{2}/\binom{n}{2}}.
\end{aligned}
$$

It follows that

$$
\begin{aligned}
\sum_{s=0}^{S} E(I_s) &\leq \sum_{s=0}^{S}\left(\frac{en}{q}\right)^q \left(\frac{eq^2}{2}\right)^s e^{-m\binom{q}{2}/\binom{n}{2}} \\
&\leq \left(\frac{en}{q}\right)^q \left(\frac{eq^2}{2}\right)^{S+1} e^{-m\binom{q}{2}/\binom{n}{2}} \\
&= e^{q(1+\ln n - \ln q)+(S+1)(1+2\ln q)-mq(q-1)/(n(n-1))}.
\end{aligned}
$$

Now as

$$
\begin{aligned}
&q(1 + \ln n - \ln q) + (S+1)(1 + 2\ln q) - mq(q-1)/(n(n-1)) \\
&< \frac{n}{k}(1 + \ln n) + \left(1 + \frac{(5k^3)^g}{3}\right)(1 + 2\ln n) - kn/2 \qquad \to -\infty
\end{aligned}
$$

we have that

$$\lim_{n\to\infty}\sum_{s=0}^{S} E(I_s) = 0.$$

In particular, if E_2 is the event that there is *no* set of $q = n/k$ vertices having at most $S = \dfrac{(5k^3)^g}{3}$ edges, then for n sufficiently large,

$$\text{Prob}(E_2) > \frac{1}{2}. \tag{3.11}$$

We deduce from (3.10) and (3.11) that

$$\text{Prob}(E_1 \cap E_2) > 0$$

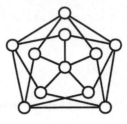

FIGURE 3.16: The Grötzsch graph

so there is a graph G that has at most $S = \dfrac{(5k^3)^g}{3}$ cycles of length less than g and for which no set of $q = n/k$ vertices has at most $S = \dfrac{(5k^3)^g}{3}$ edges. For this graph, delete one edge from each cycle of length less than g; it is not hard to verify (see Exercise 3.40) that the resulting graph is not k–colourable, and by construction, it has no cycles of length less than g.   ∎

The **girth** is defined to be the length of the shortest cycle (it is set to ∞ if a graph is **acyclic**, that is, has no cycles). It is easy to deduce from Theorem 3.16 that for all positive integers k, $g \geq 3$ there is a k–chromatic graph with girth g. There are some constructions (mostly algebraic) of graphs of high chromatic number without small cycles but all are fairly complex. The smallest 4–chromatic triangle-free graph (known as the **Grötzsch graph**) is shown in Figure 3.2.5.

Exercises

Exercise 3.1 *Show that the Laplacian of the complete graph K_n is $nI_n - J_n$, and hence find its spectrum.*

Exercise 3.2 *Prove that indeed the Laplacian of a loopless multigraph G is positive semidefinite. (Hint: show that for any vector $\mathbf{v} \in \mathbb{R}^n$, $\mathbf{v}^{\mathrm{T}} L\mathbf{v} \geq 0$.)*

Exercise 3.3 *Use the Matrix Tree Theorem to prove that the complete graph K_n of order n has exactly n^{n-2} spanning trees.*

Exercise 3.4 *Explain how the Matrix Tree Theorem can be used to count the number of spanning trees in polynomial time.*

Exercise 3.5 *What happens to the number of spanning trees in a graph if every edge of the graph is replaced by k edges in parallel? Explain your answer both with and without the Matrix Tree Theorem.*

Exercise 3.6 *For directed graph D with $n \times n$ adjacency matrix A, prove that D is strongly connected if and only if $I + A + A^2 + \cdots + A^{n-1}$ has all nonzero entries.*

Exercise 3.7 *Prove Theorem 3.3.*

Exercise 3.8 *Show that any critical graph is vertex critical.*

Exercise 3.9 *Show that any k–chromatic graph $(k \geq 2)$ contains a k–critical subgraph.*

Exercise 3.10 *Show that any k–critical graph has minimum degree at least $k - 1$.*

Exercise 3.11 *Show that the vertex 3–critical graphs are precisely the odd cycles.*

Exercise 3.12 *Show how one can determine in polynomial time for any graph G whether G is 2–colourable.*

Exercise 3.13 *Find a tiling of the plane with regular hexagons that shows that $\chi(\mathbb{G}^2) \leq 7$.*

Exercise 3.14 *Find a set of seven points in the plane that shows that $\chi(\mathbb{G}^2) \geq 4$.*

Exercise 3.15 *Find a recurrence for the reliability of a cycle of order n from the Factor Theorem.*

Exercise 3.16 *Find the reliability of the following graph.*

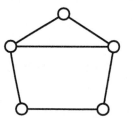

Exercise 3.17 *Find the reliability of the following multigraph.*

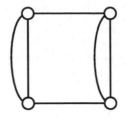

Exercise 3.18 *Prove the recursion formula*

$$Rel(K_n, p) = 1 - \left(\sum_{i=1}^{n-1} \binom{n-1}{i-1} (1-p)^{(n-i)i} Rel(K_i, p) \right).$$

(Hint: Consider in a subgraph of K_n the component vertex 1 lies in.)

Exercise 3.19 *Find a recursive formula for $Rel(K_{m,n}, p)$.*

Exercise 3.20 *Show that for any loopless multigraph of order n and size m that $F_i = 0$ for $i > m - n + 1$ and that F_{m-n+1} counts the number of spanning trees in the graph. Deduce that one can calculate F_{m-n+1} in polynomial time.*

Exercise 3.21 *Assume that the coefficients in the H-form of the reliability polynomial are always positive integers (we shall prove as much in Chapter 6). Prove Theorem 3.5.*

Exercise 3.22 *Explain the derivation of (3.2).*

Exercise 3.23 Hall's Theorem *states that in a bipartite graph G with bipartition (X, Y), there is a matching that meets every vertex of X if and only if for every subset S of X, $|N_G(S)| \geq |S|$ (a **matching** is a subset of edges that share no end points with each other, and the **neighbourhood** $N_G(S)$ of S is defined $\{w : \{w, s\} \in E$ for some $s \in S\}$). A **doubly stochastic matrix** is a square nonnegative matrix such that every row and column adds up to 1. Prove from Hall's Theorem that every doubly stochastic matrix M can be written as the convex combination of permutation matrices, that is, there are permutation matrices $P_1, \ldots, P_k$ and real numbers $\lambda_1, \ldots, \lambda_k$, all in $(0, 1]$ such that $M = \lambda_1 P_1 + \cdots \lambda_k P_k$ and $\lambda_1 + \ldots + \lambda_k = 1$.*

Exercise 3.24 *Deduce from Hall's Theorem that if H is a subgroup of order l of a finite group $(G, *)$ with index k (i.e. there are exactly k left and right cosets of H in G) then there are distinct elements $g_1, \ldots, g_k$ of G such that $g_1 * H, \ldots, g_k * H$ are the left cosets of H and $H * g_1, \ldots, H * g_k$ are the right cosets of H.*

Exercise 3.25 *Let $\beta(G)$ denote the* **independence number** *of G, that is, the order of the largest independent set of G. For a graph G of order n, what is the relationship between $\tau(G)$ and $\beta(G)$?*

Exercise 3.26 *Prove that in $\mathbb{R}^t$ and any $k \geq 1$, one can choose k vectors in general position. (Hint: The measure of any finite number of hyperplanes of dimension $t - 1$ is always 0.)*

Exercise 3.27 *Prove that the vector space $Symm_t$ of all symmetric $t \times t$ matrices has dimension $\binom{t+1}{2}$.*

Exercise 3.28 *In the proof of Theorem 3.7, explain why there is a nonzero vector $\mathbf{c} \in \mathbb{R}^t$ orthogonal to the set $\{\mathbf{b} : b \in B\}$.*

Exercise 3.29 *In the proof of Theorem 3.7, explain why $\{\mathbf{b} : b \in B\}$ is a basis for the orthogonal complement of $Span(\mathbf{c})$, and why this implies that neither $\mathbf{u}$ nor $\mathbf{v}$ is orthogonal to $\mathbf{c}$.*

Exercise 3.30 *Prove the following result of Tverberg [99]: The edges of K_n cannot be partitioned into at most $n - 2$ complete bipartite graphs.*

Exercise 3.31 *Suppose that $h(x)$ is a nonzero polynomial of degree d with real coefficients. Prove that $h(x)$ has all real roots iff $H(x) = x^d \cdot h(1/x)$ does.*

Exercise 3.32 *Provide a very short and tidy proof that for any $n \geq 1$, the binomial sequence $\langle \binom{n}{0}, \binom{n}{1}, \ldots, \binom{n}{n} \rangle$ is strictly log concave.*

Exercise 3.33 *Prove, with the notation of Theorem 3.12, that*

$$Rel(G_l, p) = lp^{l-1}(1 - p) Rel((G - e)', p) + p^l Rel(F, p).$$

Exercise 3.34 *Prove that one can choose l large enough in the proof of the previous theorem that the roots of $Rel(G_l, p)$ are as close as one would like to the roots of $Rel((G - e)', p)$.*

Exercise 3.35 *Prove that if the standard form of the reliability polynomial has all real roots, then so do $\sum F_i x^i$, $\sum N_i x^i$ and $\sum H_i x^i$. Conclude that if the reliability polynomial has all real roots, then all of the sequences associated with reliability are log concave.*

Exercise 3.36 *Prove that if G' is the graph formed from G by a parallel operation, replacing edge $e = xy$ by the edges $e_1 = xy$ and $e_2 = xy$, and G'' is the graph formed from G by a series operation, replacing edge $e = xy$ by the edges $e_1 = xz$ and $e_2 = zy$ (z a new vertex), then*

$$
\begin{aligned}
Rel(G') &= Rel(G, p_e \leftarrow 1 - (1 - p_1)(1 - p_2)), \\
Rel(G'') &= (1 - (1 - p_1)(1 - p_2))\, Rel(G, p_e \leftarrow \tfrac{p_1 p_2}{p_1 + p_2 - p_1 p_2}).
\end{aligned}
$$

Exercise 3.37 *Prove that if $|1 - p_i| > 1$ for $i = 1$ and 2, then $\left|1 - \frac{p_1 p_2}{p_1 + p_2 - p_1 p_2}\right| > 1$ as well. (Hint: Consider $1/v = (1/p) - 1$ and show first that $|1 - p| > 1$ if and only if $Re(1/v) < -1/2$. Then show that if we set $1/v_e = (1/p_e) - 1$, $1/v_1 = (1/p_1) - 1$ and $1/v_2 = (1/p_2) - 1$, then $1/v_e = 1/v_1 + 1/v_2$.)*

Exercise 3.38 *Prove that if $\chi(G) = k$ for an infinite graph G, then G contains as an induced subgraph a vertex k–critical graph.*

Exercise 3.39 *Prove that no infinite graph is vertex critical.*

Exercise 3.40 *Explain why, at the end of the proof of Theorem 3.16, the graph resulting by delete one edge from each cycle of length less than g is not k–colourable.*

Exercise 3.41 *A **tournament** is a digraph $T = (V, A)$ such that for all $u, v \in V$, $u \neq v$ exactly one of (u, v) and (v, u) is an arc. Let the property T_k*

be that for any subset U of k vertices of a tournament, there is a vertex v that 'beats' all of U, that is, $(v, u) \in A$ for all $u \in U$. Prove that for every k there is a tournament with property T_k.

Exercise 3.42 *Show that each P_k in section 2.4.4 can be expressed as a sentence ϕ_k in the first-order theory of graphs, and prove that $\Phi = \{\phi_k : k \geq 1\}$ is consistent. (Hint: use Theorem 2.1 and the compactness principle.)*

Exercise 3.43 *Show that any two countable graphs, each having property P_k for all $k \geq 1$, are infinite and isomorphic.*

Exercise 3.44 *The* **Löwenheim–Skolem Theorem** *states that any countable theory, i.e. one with a countable number of predicates and axioms (sentences), that has a model has a countable model. Show that if any two models of a countable theory T are isomorphic and there are no finite models of T then the theory is* **complete**, *that is, for any sentence ρ, one can prove neither ρ nor $\neg\rho$ from $\mathcal{T}$. Conclude that Φ is complete.*

Exercise 3.45 *Derive the* **0–1 Law for Simple Graphs**: *For any first-order sentence ψ in the theory of graphs, either*

$$\lim_{n \to \infty} Prob(G \text{ satisfies } \psi) = 0$$

or

$$\lim_{n \to \infty} Prob(G \text{ satisfies } \psi) = 1.$$

Exercise 3.46 *Find a property of graphs for which the 0–1 law does not hold (of course, this will be a property not expressible in the first-order theory of graphs).*

Exercise 3.47 *Prove that the Grötzsch graph (see Figure 3.2.5) is 4-chromatic.*

Chapter 4

Preorders and Partial Orders

Order is an important notion that permeates much of mathematics. We are well acquainted with the linear orders such as the reals, rationals and integers, and anyone who has taken a course in set theory knows about how important ordinals are to the foundations of mathematics. Preorders and partial orders form a useful underlying structure for many mathematical objects.

Recall that a **preorder** $P = (V, \preceq)$ is a discrete structure with $\preceq$ being reflexive and transitive; a partial order is a preorder whose relation is also asymmetric. We define the **condensation** of a preorder P to be the preorder $\rho(P)$ whose points are the strongly connected components $S_1, \ldots, S_r$ of P, and whose arcs are

$$\{(S_i, S_j) : \exists v_i \in S_i, v_j \in S_j \text{ with } (v_i, v_j) \in \preceq\}.$$

It is not hard to see that $\rho(P)$ is asymmetric, and hence a partial order, and indeed every preorder arises from a unique partial order by replacing vertices by equivalence relations.

Example: $(\mathbb{Z}, |)$, where $\mathbb{Z}$ is the set of integers and $|$ denotes the divisibility relation is a preorder. Its condensation is isomorphic to $(\mathbb{Z}_{\geq 0}, |)$, as its strongly connected components are $\{0\}$ and sets of the form $\{-i, i\}$ for $i \geq 1$. $\qquad \triangle$

The **transitive closure** R^* of a relation R on a set V is the smallest relation R' on V that is transitive (one can form the transitive closure on a finite set by successively adding in arcs (x, z) for which (x, y), (y, z) are present but (x, z) is not).

Example: For the relation $R = \{(1, 2), (1, 3), (2, 2), (2, 4), (3, 5), (4, 5)\}$ on set

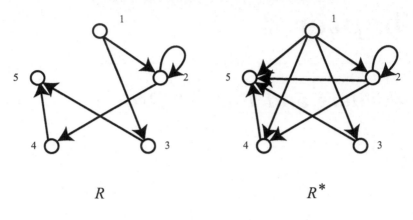

R R^*

FIGURE 4.1: A relation and its transitive closure

$X = [5]$, its transitive closure is $R^* = \{(1,2),(1,3),(1,4),(1,5),(2,2),(2,4),$
$(2,5),(3,5),(4,5)\}$ (see Figure 4.1). $\triangle$

An element $x \in V$ is **maximal** if there is no $y \in V$, $y \neq x$ such that
$x \preceq y$; x is **minimal** if there is no $y \in V$, $y \neq x$ such that $y \preceq x$. An element
$x \in V$ is a **maximum** if $y \preceq x$ for all $y \in V$; a **minimum** if $x \preceq y$ for all
$y \in V$. A partial order with both a maximum and minimum is said to be
bounded. Elements x and y are **comparable** if $x \preceq y$ or $y \preceq x$, and are
incomparable, written $x||y$, otherwise. For $v \in V$, $Inc(v) = \{y \in V : x||y\}$
and $\mathrm{Inc}(P) = \{(u,v) \in V \times V : u||v\}$ denotes the set of all incomparable pairs
in P. A **chain** in P is a subset C of V such that any two elements of C are
comparable; an **antichain** is a subset A of V such that every two elements
of A are incomparable. The **height** of a partial order and **width** of a partial
order are, respectively, the number of vertices in a maximum size chain and
antichain.

Example: In the example $(\mathbb{Z}, |)$, 0 is a maximum element. In the partial order
$(\{z \in \mathbb{Z} : z \geq 2\}, |)$, there are no maximal elements, and the minimal elements
are precisely the prime numbers. The numbers are 6 and 12 are comparable,
while 6 and 8 are incomparable. $\triangle$

Example: The partial order $B_n = (\mathcal{P}([n]), \subseteq)$ contains all subsets of $[n]$), and has height $n+1$ and width $\binom{n}{\lfloor n/2 \rfloor}$ (see Exercise 4.21). It is called a **boolean lattice**. △

Let $P = (V, \preceq)$ be a partial order. Let $S \subseteq V$. An **upper bound** for S is an element $x \in V$ such that $s \preceq x$ for all $s \in S$; x is a **lower bound** for S if $x \preceq s$ for all $s \in S$. A **least upper bound** for S, written as $\bigvee S$, is a minimum element among the upper bounds for S; a **greatest lower bound** for S, written as $\bigwedge S$, is a maximum element among the lower bounds for S. A **lattice** is a partial order such that $\bigwedge S$ and $\bigvee S$ exist for every set S of two elements (if $S = \{x, y\}$, we write $\wedge S$ as $x \wedge y$, the **meet** of x and y, and write $\vee S$ as $x \vee y$, the **join** of x and y.

Example: For any two elements X and Y in the partial order B_n, we have that $X \vee Y = X \cup Y$ and $X \wedge Y = X \cap Y$, so indeed B_n is a lattice (in fact for *any* subset S of points of B_n, $\bigvee S = \bigcup_{X \in S} X$ and $\bigwedge S = \bigcap_{X \in S} X$ exist). △

There are some other special types of preorders and partial orders worth noting.

- An **indiscrete preorder** on set X has all n^2 arcs. We denote the indiscrete preorder on $[n]$ by Indisc_n.

- An **discrete preorder** on set X has no arcs except for loops (it is, of course, a partial order as well). We denote the discrete preorder on $[n]$ by Disc_n.

- **Linear orders** are partial orders such that every two elements x, $y \in V$ are comparable, that is, $x \preceq y$ or $y \preceq x$.

A basic but important result is that every partial order $P = (X, \preceq)$ on a set X has a **linear extension**, that is, a linear order on X that contains $\preceq$ (see Exercise 4.24).

Let $P = (V, R)$ be a preorder. An **ideal** $I \subseteq V$ of P is a downwards–closed set; that is, if $(x, y) \in R$ and $y \in I$ then $x \in I$. For $v \in P$, we write $D(v)$ for the ideal generated by v, that is, $D(v) = \{u : (u, v) \in R\}$. A **filter** of a

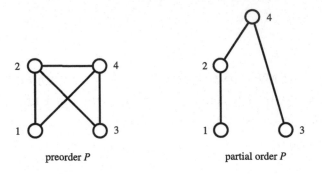

FIGURE 4.2: A Hasse and generalized Hasse diagram

preorder P is an upwards–closed set. For $v \in P$, we write $U(v)$ for the filter generated by v, that is, $D(v) = \{u : (v, u) \in R\}$.

Example: $(\mathbb{Z}_{\geq 0}, |)$, for any subset $S \subseteq \mathbb{Z}_{\geq 0}$, the set of all positive divisors of any elements of S is an ideal, and the set of multiples of any elements of S is a filter. $\triangle$

Preorders and partial orders can be represented in the plane as digraphs. On the other hand, not all of the arcs are necessary as we know the relation is reflexive and transitive. We say that y **covers** x in partial order $P = (V, \preceq)$ if $x \neq y$, $x \preceq y$ and for no $z \neq x$, y is $x \preceq z$ and $z \preceq y$. The **covering graph** $\mathrm{Cov}(P)$ of a partial order $P = (V, \preceq)$ (a model of a partial order with a graph) is the graph on V whose edges are $\{x, y\}$ such that x covers y or y covers x. A **Hasse diagram** of P is a representation of $\mathrm{Cov}(P)$ in the plane such that if $x \preceq y$ then the image of x is placed below the image of y. One can extend a Hasse diagram to a preorder by drawing the Hasse diagram of the condensation and then replaying each vertex by the corresponding equivalence class, drawn horizontally with a straight line through them (we call this the **generalized Hasse diagram** of the preorder). For an example, see Figure 4.2. The Hasse diagram for B_3 is shown in Figure 4.3.

For the readers who are interested in pursuing partial and linear orders further, we direct them to the excellent reference [97].

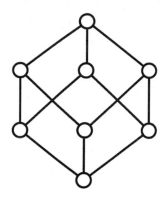

FIGURE 4.3: The Hasse diagram of B_3

4.1 Finite Topologies and Preorders

4.1.1 The Correspondence

A **topology** τ on set V is a collection $\mathcal{O}$ of subsets of V that contains $\emptyset$ and V, and is closed under arbitrary unions and finite intersections; the sets in $\mathcal{O}$ are called **open sets** of the topology; if S is an open set, then $V - S$ is called a **closed set** of τ. Note that, in fact, every topology is a certain type of hypergraph. A **basis** $\mathcal{B}$ for a topology τ is a collection of open sets such that every open set on τ is a union of elements of $\mathcal{B}$.

Example: Consider the set $V = \{1, 2, 3, 4\}$. If we take as open sets $\emptyset$, V, $\{1, 2\}, \{1, 2, 3\}$, $\{1, 2, 3, 4\}$, then it is not hard to check that the open sets are closed under arbitrary unions and finite intersections.   △

Topologies have a long history of study, arising from geometry and analysis. The definitions of a topology described here arise in **point set topology**, and describe topologies in their most abstract setting. Without a doubt the lion's share of work has been on infinite topologies, but finite topologies have had their uses, in areas such as image analysis.

It is well known (see [40]) that there is a 1-1 correspondence between topologies on $[n]$ and preorders on $[n]$ as follows. Given a topology τ on $[n]$, form the directed graph $D(\tau)$ on $[n]$ with arcs $\{(y, x) : y$ is in every open set of τ that contains $x\}$; it is trivial to see that $D(\tau)$ is reflexive and transitive. Conversely, given any reflexive, transitive directed graph D on $[n]$, one can form the topology $\tau(D)$ on $[n]$ by taking the sets $\mathcal{O}_x = \{y : (y, x)$ is an edge of $D\}$, as x ranges over $[n]$, as a basis.

Example: Consider the topology τ on $V = \{1, 2, 3, 4\}$ with open sets $\emptyset$, V, $\{1, 2\}$, $\{1, 2, 3\}$, $\{1, 2, 3, 4\}$ (see the previous example). Then $D(\tau)$ has arcs $\{(1, 1), (2, 2), (3, 3), (4, 4), (1, 2), (2, 1), (1, 3), (2, 3), (1, 4), (2, 4), (3, 4)\}$, and is, in fact, a preorder. The sets $\mathcal{O}_1 = \mathcal{O}_2 = \{1, 2\}$, $\mathcal{O}_3 = \{1, 2, 3\}$ and $\mathcal{O}_4 = \{1, 2, 3, 4\}$ form a basis for τ. $\triangle$

The connections between finite topologies and preorder have been utilized in enumeration and other problems on finite topologies [40, 55, 85, 86].

The following observations will be of use to us.

Observation 4.1 *For any two topologies τ and σ on finite sets X and Y respectively, a map $f : X \to Y$ is continuous iff f is an* **ordering preserving map** *between the preorders $D(\tau) = (X, \preceq_X)$ and $D(\sigma) = (Y, \preceq_Y)$, that is, $x_1 \preceq_X x_2$ implies that $f(x_1) \preceq_Y f(x_2)$ for all $x_1, x_2 \in X$.*

Observation 4.2 *For any two topologies τ and σ of order n, τ and σ are* **homeomorphic** *(i.e. there is a bijection $f : V(\tau) \to V(\sigma)$ such that both f and f^{-1} are continuous) if and only if $D(\tau)$ and $D(\sigma)$ are isomorphic (as directed graphs).*

Observation 4.3 *For any two topologies τ and σ of order n, $E(\tau) \subseteq E(\sigma)$ if and only if $A(D(\sigma)) \subseteq A(D(\tau))$.*

Observation 4.4 *The open sets of a topology τ on a finite set V are precisely the ideals in the preorder $D(\tau)$, and the closed sets are precisely the filters of $D(\tau)$.*

Observation 4.5 *The indiscrete topology on a finite set V (whose only open sets are $\emptyset$ and V) corresponds to the indiscrete preorder on V. The discrete* *topology on a finite set V (whose open sets are all subsets of V) corresponds to the discrete preorder on* $[n]$.

Observation 4.6 *A* **T$_0$ topology** *on a set X is one that satisfies the T_0 separation axiom: for every x and y, there is either an open set containing x and not y, or vice versa. A finite topology τ is T_0 iff $D(\tau)$ is a partial order.*

We will often identify τ and $D(\tau)$, and hence discuss such things as "τ and σ are arc disjoint", "the transitive closure of $\tau \cup \sigma$" or "a maximal element of τ."

4.1.2   Open Sets

We have seen that open sets of a finite topology correspond to ideals in the corresponding preorder. This correspondence can help us in answering various questions about open sets in a finite topology. For example, how many open sets can a topology on a set V of cardinality n have? It certainly has at least 2, namely $\emptyset$ and V, as the indiscrete topology on V has, and can have at most 2^n, as the discrete topology on V has. Can we find a topology on V with l open sets for any $2 \leq l \leq 2^n$? The answer is no, and the proof given via preorders in shorter than the original argument [91].

Proposition 4.7 *If τ is a topology on a set V of size n and τ is not discrete, then τ has at most $3 \cdot 2^{n-2}$ many open sets.*

Proof: We turn the problem into counting the number of ideals in a non-discrete preorder on V. As the preorder $D(\tau)$ is not discrete, there is an arc (x, y) in the preorder with $x \neq y$. Consider any ideal that contains y; it necessarily contains x, hence there are at most 2^{n-2} of these, as there are 2^{n-2} subsets of $V - \{x, y\}$. The number of ideals that do not contain y is at most 2^{n-1}. Thus there are at most $2^{n-1} + 2^{n-2} = 3 \cdot 2^{n-2}$ ideals, and it follows that τ has at most $3 \cdot 2^{n-2}$ open sets. ∎

This shows that there can be big gaps in the spectrum of the number of open sets of a finite topology.

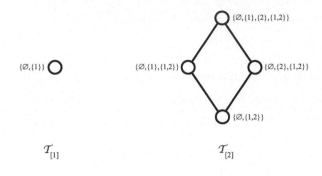

FIGURE 4.4: Lattice of topologies on sets of orders 1 and 2

4.1.3   The Lattice of All Topologies

Let Top_V denote all topologies on set V. We make Top_V into a partial order $\mathcal{T}_V = (\mathrm{Top}_V, \subseteq)$: $\tau \subseteq \sigma$ if and only if the open sets of τ are open in σ. This makes $(\mathrm{Top}_V, \subseteq)$ into a lattice, as the largest topology contained in τ and σ is the topology on V whose open sets are open in both τ and σ, while the smallest topology containing τ and σ is the topology on V whose open sets are open in each topology that contains both τ and σ (this set is nonempty as it contains the discrete topology on V). (For more information on this lattice, see Larson and Andima's survey [59].) The lattices of topologies on sets of size 1 and 2 are shown in Figure 4.4.

One can also make the set of preorders on V a lattice in the same way by defining $P = (V, R) \leq P'(V, R')$ if and only if $R' \subseteq R$. Let us call this lattice Pre_V. By Observation 4.3, Pre_V is isomorphic (as a partial order) to Top_V.

The definition of meet and join is easier to describe in $\mathrm{Pre}_{[n]}$.

Theorem 4.8 *For any two preorders* $P = (V, R)$ *and* $Q = (V, R')$ *on* $[n]$, $P \vee Q = (V, R \cap R')$, *while* $P \wedge Q = (V, (R \cup R')^*)$ *(recall that* $*$ *denotes the transitive closure).* ∎

4.1.4   Algorithmic Considerations

One of the fascinating questions that arises on finite topologies is the computational complexity of various topological parameters. How easy is it to decide whether a finite topology is connected? How difficult is it to determine whether two finite topologies are homeomorphic? We'll look at these and other issues.

We need to discuss how to machine represent finite topologies. The usual way to represent a topology from a topologist's point of view is to simply list its open sets (the **open set representation**). However, this will not suffice for serious computation, as a topology on n points may have 2^n sets. In particular, the discrete topology of order 20 has over one million open sets, but clearly we do not need all of these to identify the topology.

Under the representation of a finite topology by a preorder, a topology on n points has size $O(n^2)$ (note that one cannot hope for a substantially better representation as we shall see that there are

$$2^{\frac{n^2}{4}+O(n)}$$

many topologies on n points, and if we represent topologies in a finite alphabet, at least one of these will have input length at least a constant times n^2). This representation is equivalent to a listing of the minimal open sets around each point, and is a compact representation of the topology. If one wants to further reduce the size of the input, one can represent the topology by a listing its generalized Hasse diagram. For most machine representations, the adjacency matrix of the preorder has been the most frequent model.

One can inductively generate the adjacency matrices of all preorders on $[n]$ by the following algorithm [40]. It will be easiest to generate the transitive irreflexive relations, and then add on the loops at each vertex. Suppose we have already generated the adjacency matrices of all transitive irreflexive relations of order $n - 1$, say as $A_1, \ldots, A_l$. For any such matrix $A = A_i$ we now form all block matrices of the form

$$B = \begin{pmatrix} A & \alpha^T \\ \beta & 0 \end{pmatrix}$$

where α, $\beta \in \mathbb{R}^{n-1}$. To ensure that B is the adjacency matrix of a transitive irreflexive relation, we need only ensure the following conditions:

1. $a_{i,j} = 0 \Rightarrow \alpha_i \beta_j = 0$ for all $1 \leq i,\ j \leq n-1$

2. $\alpha_i = 0 \Rightarrow a_{i,j} \alpha_j = 0$ for all $1 \leq i,\ j \leq n-1$

3. $\beta_j = 0 \Rightarrow a_{i,j} \beta_i = 0$ for all $1 \leq i,\ j \leq n-1$

Once you have completed the list of all such B, simply add 1s to the diagonal to get the desired adjacency matrices of all preorders of order n. If one wishes to generate all partial orders of order n inductively, one starts with a list of all acyclic transitive relations of order $n-1$, chooses A from this list, ensures that the above 3 conditions hold, and that the resulting matrix B corresponds to an acyclic relation. To check the latter, note that a directed graph is acyclic if and only if it always has a vertex of out-degree 0 (and a vertex of in-degree 0), so you can recursively strip away such vertices until you either have a singleton (in which case the original directed graph was acyclic) or not (in which case the original directed graph was not acyclic). Again, once you have completed the list of all such B, simply add 1s to the diagonal to get the desired adjacency matrices of all partial orders of order n.

From a complexity point of view, the sizes of the preorder and generalized Hasse diagrams are polynomially related, and hence the existence of polynomial time algorithms for one is equivalent to the other. Such is not the case for the 'open set representation', as the size of the input is rather inflated. For example, it is trivial to count the number of open sets of a finite topology given the open set representation, but we shall see that counting the number of open sets for the preorder (or generalized Hasse diagram) representation is intractable.

Note that determining whether a subset S of points in a topology on n points is open can be carried out in $O(|S|n)$ time, as one need only check for each vertex in S whether all of its in–neighbourhoods are contained in S. Similarly, the open set generated by a subset S can be determined in $O(|S|n)$ time, as it is simply the union of all the in–neighbourhoods of points of S. We will need these two observations from time to time.

One property of a topological space that one may wish to determine is that of connectedness (a space is disconnected if it is the disjoint union of two proper open sets). On the surface, it is apparent that connectedness is in co–NP, as one need only provide the two disjoint proper open sets whose union is the whole space. However, much more is true.

Theorem 4.9 *The problem*

INPUT: Topology σ (in a compact representation).

QUESTION: Is σ connected?

is polynomial.

Proof: Note that a topology is connected if and only if the associated preorder is weakly connected as a directed graph. To check this, we can search for a spanning tree in the underlying graph, and this can be done in linear $(O(n))$ time from the generalized Hasse diagram. ∎

We now turn to the difficulty of determining if two topologies are isomorphic. Let's state the general problem as follows:

HOMEOMORPHISM

INPUT: Topologies σ and τ (in a compact representation).

PROBLEM: Are σ and τ homeomorphic?

The inherent difficulty is that the obvious algorithm, of checking whether there exists a continuous bijection, requires considering on the order of $n!$ maps. It is clear that the problem is in **NP** as given a prospective map, one can verify quickly that it is a bijection and that it is continuous (recall that a function between topological spaces is continuous if and only if it is an order preserving map between the associated preorders, and hence continuity between two topological spaces on n points can be verified in $O(n^2)$ time).

We note that a similar problem for graphs, namely graph isomorphism, is not known to be NP–complete. In fact, the complexity of graph isomorphism is one of the outstanding open problems in theoretical computer science, and has withstood considerable scrutiny. As a result, a class of problems has arisen, known as **isomorphism complete**, being those problems for which if any of them had a polynomial time algorithm, then graph isomorphism would (note the analogy to NP–complete problems). Many problems in graph theory are known to be isomorphism complete , and hence very likely to be intractable. Franklin and Zalcstein [45] have shown the following:

Theorem 4.10 *HOMEOMORPHISM is isomorphism complete.*

Proof: Let σ and τ be two topologies. They are homeomorphic if and only if their corresponding preorders $P(\sigma)$ and $P(\tau)$ are isomorphic, and thus the problem is equivalent to showing that preorder isomorphism is isomorphism complete.

Given a graph G, we form a preorder $P(G)$ on $V(G) \cup E(G)$ with arcs (v, e) where v is an end of e. It is clear that $P(G)$ can be constructed from G in polynomial time. Moreover, G can be quickly reconstructed from $P(G)$, as the edges of G are the vertices with two incoming arcs, from its endpoints. Thus given two graphs G and H, we can construct $P(G)$ and $P(H)$ in polynomial time, and moreover $G \cong H$ if and only if $P(G) \cong P(H)$, i.e. if and only if the topologies corresponding to $P(G)$ and $P(H)$ are homeomorphic. The isomorphism completeness of HOMEOMORPHISM follows. ∎

Counting the number of open sets for a topology is trivial if the topology is given in the bloated open set representation. However, the problem of counting the number of open sets becomes much more interesting when considering the compact preorder model. There, open sets correspond to ideals. A topology on n points can have as many as 2^n (as the discrete topology shows), so listing all open sets is not a viable method for counting them.

It is not hard to see that if P is a preorder with condensation P' (which is a poset), then the number of open sets in P and P' are identical, and, moreover, this number is equal to the number of antichains in P', since the antichain of the maximal elements of an open set completely determined the open set. Now Ball and Provan [5] have shown that counting the number of antichains in a poset is $\#P$–complete, and hence we have:

Theorem 4.11 *The problem*

    INPUT: Topology τ

    QUESTION: How many open sets does τ have?

is #P–complete. ■

An immediate corollary is

Corollary 4.12 *The problem*

    INPUT: Topology τ

    QUESTION: How many closed sets does τ have?

is #P–complete. ■

4.2   Representing Preorders and Partial Orders

4.2.1   Random Preorders and Partial Orders

Given a property of preorders or partial orders, a natural question to ask about preorders and partial orders is whether most preorders and partial orders have or don't have the property. And this brings to mind two questions:

- How many preorders and partial orders are there on a set of cardinality n?

- What do "most" preorders and partial orders look like?

Table 4.1 shows the number of topologies, and the number of T_0 topologies, on a set of size at 13, and there are no discernible patterns. In fact, there is no nice closed formula for the first question; the best answer we have is a difficult asymptotic result due to Kleitman and Rothschild [55].

Theorem 4.13 ([55]) *The number of partial orders on a set of size n is* $2^{n^2/4+O(n^{3/2}\ln n)}$.

n	number of topologies on $[n]$	number of T_0-topologies on $[n]$
1	1	1
2	4	3
3	29	19
4	355	219
5	6,942	4,231
6	209,527	130,023
7	9,535,241	6,129,859
8	642,779,354	431,723,379
9	63,260,289,423	44,511,042,511
10	8,977,053,873,043	6,611,065,248,783

TABLE 4.1: Number of topologies on a finite set

What is more surprising than the actual count is what most partial orders look like. You might expect most partial orders to have height and width about $\sqrt{n}$, but you would be wrong! It was shown in [55] that the Hasse diagram of almost all partially ordered sets of order n consists of three levels L_1, L_2 and L_3, with L_1 the minimal elements, L_3 the maximal elements, and $|L_1|$, $|L_3| = \frac{n}{4} + o(n)$. For every such choice of three sets L_1, L_2 and L_3 we make this into a probabilistic model by covering $x \in L_i$ with $y \in L_{i+1}$ ($i = 1$ or 2) with probability $1/2$. By $\Omega_n(p, q, r)$ we denote the sample space of all such partial orders on $[n]$ with $|L_1| = p$, $|L_2| = q$ and $|L_3| = r$ formed by taking a random ordered partition of $[n]$ into sets of these cardinalities and then choosing the edges as discussed. We let $\Omega_n = \bigcup_{p,q,r} \Omega_n(p, q, r)$, with the union over all p, q and r with p, $r = n/4 + o(n)$ and $q = n/2 + o(n)$ (and call any such poset a **KR poset**).

As an example of how one might use this model, consider the following:

Lemma 4.14 ([24]) *Almost all partial orders P on n vertices have every element of L_3 above every element of L_1.*

Proof: It suffices to show this for each $\Omega_n(p, q, r)$. If some $x \in L_1$ is not below some $y \in L_3$, then for the subset S of L_2 that covers x, Y does not cover any

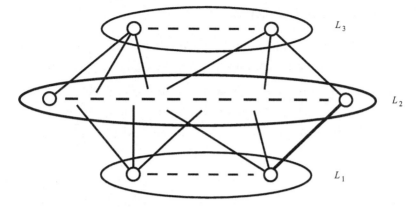

FIGURE 4.5: A random poset "sandwich" (with a lot of meat in the center)

element of S. It follows that the probability that some $x \in L_1$ is not below some $y \in L_3$ is at most

$$pr \sum_{i=0}^{q} \binom{q}{i} \left(\frac{1}{2}\right)^i \left(\frac{1}{2}\right)^{q-i} \left(\frac{1}{2}\right)^i = pr \left(\frac{3}{4}\right)^q \leq \frac{n}{15} \left(\frac{3}{4}\right)^{n/3}.$$

As $\Omega_n = \bigcup_{p,q,r} \Omega_n(p,q,r)$ we have that in Ω_n the probability that some $x \in L_1$ is not below some $y \in L_3$ is at most

$$\frac{n}{15} \left(\frac{3}{4}\right)^{n/3} \to 0 \text{ as } n \to \infty,$$

so it immediately follows that the probability that all $x \in L_1$ are below all $y \in L_3$ tends to 1 as $n \to \infty$. ∎

Erné [39] proved that almost every preorder is a partial order. Hence we deduce the following:

Corollary 4.15 *Almost every finite topology is connected.*

Proof: This follows as the previous lemma shows that almost every partial order of the Kleitman–Rothschild form is weakly connected. ∎

4.2.2   Graphs for Preorders

In a bounded lattice, with minimum 0 and maximum 1, we say that two elements x and y are **complements** if $x \vee y = 0$ and $x \wedge y = 1$; the lattice is said to be **complemented** if every element has a complement. For example, any boolean lattice is complemented, as for any subset A of a set X, its complement in $(\mathcal{P}(X), \subseteq)$ is clearly $X - A$ (the 0 and 1 are respectively and X, and $\wedge = \cap$ while $\vee = \cup$).

Recall that the preorders on a set X form a complete bounded lattice under containment. How does one go about studying complementation on this lattice? Well, clearly the property of being a complement is symmetric, so it defines a graph on the set of all preorders (or on the set of all topologies) on a finite set V. Hence we can investigate graph–theoretic properties of the **complementation graph** on V as a means to say some interesting things about the nature of the complementation relation on the class of all preorders and topologies on V.

The study of complementation began with Hartmanis' work [52] where it was shown that every finite topology had a complement; much work has been carried out by others on the infinite case (c.f. [90, 100, 77, 78, 1, 2]). We consider here the finite case, and deal interchangeably with $\mathcal{T}_{[n]}$ and $\mathrm{Pre}_{[n]}$.

From Section 4.1.3, we see two topologies τ and σ are **complements** if $\tau \wedge \sigma$ is the indiscrete topology on X and $\tau \vee \sigma$ is the discrete topology on X. From Theorem 4.8, topologies σ and τ of order n are complementary if and only if $D(\sigma)$ and $D(\tau)$ are arc–disjoint (except for loops) and their union is strongly connected; this converts the abstract problem of complementation into a particularly useful visual one.

Example: The two preorders shown in Figure 4.6 are complementary as they are arc disjoint (except for loops) and their union is strongly connected. $\triangle$

We begin with a theorem, due to Hartmanis [52] showing that the lattice $\mathcal{T}_{[n]}$ is complemented. Our proof is shorter and uses the connection between finite topologies and preorders.

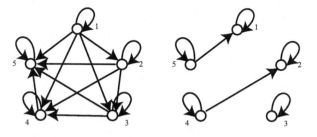

FIGURE 4.6: Two complementary preorders

Theorem 4.16 ([52]) *The lattice $\mathcal{T}_{[n]}$ is complemented, and in fact every non-discrete, nonindiscrete topology has at least 2 complements.*

Proof: It suffices to show that if $P = (V, \preceq)$ is any preorder on $[n]$ that is not discrete nor indiscrete, then it has a complement. Let $Q = \rho(P)$ be the condensation of P; it is a partial order say on set W. As P is not indiscrete, Q has at least two vertices. It suffices to show that Q has a complement, since any complement of Q can be "blown up" to a complement of P by replacing each vertex of Q by a discrete partial order on the equivalence class of that vertex in P.

If Q contains an edge (x, y), then it contains distinct elements u and v with u minimal, v maximal and $u < v$. Let L be a linear extension of Q with u at the bottom and v at the top. Then $d(L)$, the dual of L, is a complement of Q, as clearly it shares no arcs but the loops with Q, and $d(L) \cup Q$ is strongly connected as in $d(L)$ all elements are between v and u, and (u, v) is an arc of Q. (In fact, if Q is not a linear order itself, then there will be at least two such linear extensions, giving rise to at least two complements of Q).

The only remaining case is that Q is a discrete partial order on at least 2 vertices, which means that P was the disjoint union of $k \geq 2$ indiscrete preorders, with one of cardinality at least 2. Pick a point v_i $(i = 1, \ldots, k$ in each of these k parts, and form the preorder on $[n]$ that is the disjoint union of the indiscrete preorder on $\{v_i : i = 1, \ldots, k\}$ with a discrete preorder on the remaining vertices. This is a complement to P (and indeed there are at least 2 choices for such P as one component of P has at least 2 choices for a vertex to choose). ■

Let Comp_n denote the graph on $\mathrm{Top}_{[n]}$ whose edges denote complementation in $\mathcal{T}_{[n]}$. The previous result is merely a statement about the minimum degree of such a graph: $\delta(\mathrm{Comp}_n) \geq 2$ for $n \geq 2$. The best known result for $\delta(\mathrm{Comp}_n)$ is in [25], where it was shown, with a few exceptions, $\deg_{\mathrm{Comp}_n}(\tau) \geq 2^n$. Other graph–theoretic parameters that have been investigated for Comp_n include clique number [23] and diameter [24]. In [24] it was shown that in fact the complementation relation is about as varied as can be expected – every graph occurs as an induced subgraph of some Comp_n. The random partial orders were used in [24] to find upper and lower bounds on the independence number of Comp_n.

Theorem 4.17 ([24]) *If $p_n = |\mathrm{Comp}_n| = |\mathrm{Top}_{[n]}|$, then the maximum cardinality $\beta(n)$ of a complement–free subset of $\mathrm{Top}_{[n]}$ (i.e. Comp_n) is*

$$\frac{1}{4} + o(1) \leq \frac{\beta(n)}{p_n} \leq \frac{1}{2} + o(1).$$

Proof: Consider only KR partial orders with every minimal element below every maximal element and with every element in the middle layer covered above and below (we have seen that almost every KR partial order has these properties). Note that if P is any such poset then its dual $d(P)$ is a complement, as every element in the middle layer is between elements at the other two layers, and that in $P \cup d(P)$, the lower and upper layers form a strongly connected component. Thus any independent set in Comp_n can contain at most one from each such pair, so it follows (as p_n differs from $|\Omega_n|$ by at most $o(p_n)$) that

$$\frac{\beta(n)}{p_n} \leq \frac{1}{2} + o(1).$$

For the other bound, note that all KR partial orders that share a fixed minimal element, say v, are non-complementary (if $n \geq 2$), as v will be a source in the transitive closure of the union of any two such partial orders. Now the number of KR partial orders in $\Omega_n(p, q, r)$ that have v as a minimal element is

$$\binom{n-1}{p-1} \Big/ \binom{n}{p} = p/n = 1/4 + o(1)$$

so extending this to Ω_n we get the same bound. Again, as p_n differs from $|\Omega_n|$

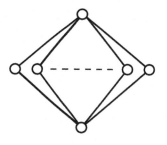

FIGURE 4.7: Conjectured preorder on $[n]$ with the most complements

by at most $o(p_n)$, we derive that

$$\frac{1}{4} + o(1) \leq \frac{\beta(n)}{p_n},$$

and we are done. ∎

We end with a conjecture.

Conjecture 4.18 ([24]) *The preorder on $[n]$ that has the most complements is the partial order with 1 above a discrete set on $\{2,\ldots,n-1\}$ which is above n.*

Exercises

Exercise 4.1 *Consider the partial order $(V, \subseteq)$ where V is the set of subsets of $[n]$ of odd size. How many minimal elements are there? How many maximal elements are there? When does this partial order have a maximum element?*

Exercise 4.2 *Prove that any finite partial order has at least one maximal element and one minimal element. What elements are both maximal and minimal?*

Exercise 4.3 *Show that any maximum in a partial order is maximal, and any minimum is minimal, and that maximums and minimums, if they exist, are unique. Do partial orders need to have maximum or minimum?*

Exercise 4.4 *Prove that the set of maximal elements and the set of minimal elements in any partial order are antichains.*

Exercise 4.5 *If $P_1 = (V_1, R_1)$ and $P_2 = (V_2, R_2)$ are disjoint preorders, then prove that the* **sum** $P_1 + P_2 = (V_1 \cup V_2, R_1 \cup R_2 \cup \{(u,v) : u \in V_1, \ v \in V_2\})$ *is a preorder, and that if P_1 and P_2 are partial orders, then so is their sum. What are the maximal and minimal elements of $P_1 + P_2$?*

Exercise 4.6 *The* **comparability graph** *of a poset $P = (V, \preceq)$ is the graph on v whose edges are $\{\{u,v\} : u \neq v, \ u \preceq v$ or $v \preceq u\}$. The* **incomparability graph** *of P is the graph on V whose edges are $\{\{u,v\} : u\|v\}$. Prove that the complement of the comparability graph of a poset P is the incomparability graph of the same poset.*

Exercise 4.7 *Do infinite partial orders need to have maximal or minimal elements?*

Exercise 4.8 *For a k–chromatic graph G, form a partial order on the k–chromatic induced subgraphs of G, ordered by vertex containment. To what elements do the vertex k–critical subgraphs belong? Conclude that every k–chromatic graph has a vertex k–critical subgraph.*

Exercise 4.9 *Find a lattice L that is not bounded.*

Exercise 4.10 *A lattice L is* **distributive** *if for any elements x, y and z we have*

$$x \vee (y \wedge z) = (x \vee y) \wedge (x \vee z)$$

and

$$x \wedge (y \vee z) = (x \wedge y) \vee (x \wedge z).$$

Prove that any boolean lattice is distributive.

Exercise 4.11 *An* **alternating cycle** *in partial order $P = (V, \preceq)$ is a list of ordered pairs $(x_0, y_0), \ldots, (x_{k-1}, y_{k-1})$ with each $x_i\|y_i$, such that $y_j \preceq x_{j+1}$ for $j = 0, \ldots, k+1$ (addition mod k); such a cycle is called* **strict** *if $y_j \preceq x_l$ if and only if $l = j + 1$. Prove that any alternating cycle contains a strict alternating cycle, and that if $(x_0, y_0), \ldots, (x_{k-1}, y_{k-1})$ is a strict alternating cycle, then (i) $\{x_0, \ldots, x_{k-1}\}$ and $\{y_0, \ldots, y_{k-1}\}$ are antichains, and (ii) $y_j\|x_l$ if and only if $l \neq j + 1$.*

Exercise 4.12 *Show that the transitive closure of a digraph with no cycles of length greater than 1 is asymmetric.*

Exercise 4.13 *Let $P = (V, \preceq)$ be a partial order. Prove that the following are equivalent for $S \subseteq Inc(P)$ [98]:*

1. *$(V, (\preceq \cup S)^*)$ is not a partial order.*

2. *S contains an alternating cycle.*

3. *S contains a strict alternating cycle.*

Exercise 4.14 *Produce an algorithm that produces for any partial order $P = (V, \preceq)$ a linear extension. Can your algorithm produce every linear extension of P?*

Exercise 4.15 *How many linear extensions do the following have? (a) A linear order on n vertices? (b) An antichain A_n on n vertices? (c) $A_n + A_m$?*

Exercise 4.16 *Prove that any partial order P is the intersection of all of its linear extensions (a set of such linear extensions is called a **realizer** of P and the minimum cardinality of a realizer is called the **dimension** of P).*

Exercise 4.17 *Let $P = (V, \preceq)$ be a finite partial order. An ordered pair (x, y) of incomparable elements is called a **critical pair** in P if $D(x) \subseteq D(y)$ and $U(y) \subseteq U(x)$. We let $Crit(P) = \{(u, v) : (u, v)$ is a critical pair of $P\}$. Prove that if $x \| y$ then there are u and v such that $u \preceq x$, $y \preceq v$ and (u, v) is a critical pair, and that if $(u, v) \in Crit(P)$, then $(V, \preceq \cup\{(u, v)\})$ is a partial order.*

Exercise 4.18 *Let $P = (V, \preceq)$ be a finite partial order. Prove a result of Rabinovitch and Rival [71] that states that a collection $\mathcal{R}$ of linear orders on V is a realizer of P if and only if $\mathcal{R}$ **reverses all critical pairs**, that is, for any $(x, y) \in Crit(P)$, there is some linear order $L = (V, \preceq_L) \in \mathcal{R}$ such that $y \preceq_L x$. Conclude that the dimension of P is the minimum cardinality of a realizer of P that reverses all critical pairs.*

Exercise 4.19 *Prove that in any finite partial order $P = (V, \preceq)$ of height h, V can be partitioned into h antichains.*

Exercise 4.20 *Prove* **Dilworth's Theorem**: *In any finite partial order $P = (V, \preceq)$ of width w, V can be partitioned into w chains.*

Exercise 4.21 *Prove that B_n has height $n + 1$ and width $\binom{n}{\lfloor n/2 \rfloor}$.*

Exercise 4.22 *Let $P = (V, R)$ be a preorder. Show that P and $\rho(P)$ have the same number of ideals.*

Exercise 4.23 *Let $P = (V, R)$ be a preorder, and let $\mathcal{I}$ be the set of all ideals of P. Show that $(\mathcal{I}, \subseteq)$ forms a complete bounded lattice.*

Exercise 4.24 *Prove that for any partial order $P = (V, \preceq)$ there is a linear order $L = (V, \preceq')$ with $\preceq \subseteq \preceq'$ (L is called a* **linear extension** *of P).*

Exercise 4.25 *Let $P = (V, R)$ be a preorder. The* **dual** *of P is the preorder $d(P) = (V, R^{-1})$.*

1. *Show that P is a partial order if and only if $d(P)$ is.*

2. *Show that if P is a partial order then P and $d(P)$ have the same dimension.*

3. *Show that L is a lattice if and only if $d(L)$ is.*

4. *Show that F is a filter in P if and only if F is an ideal in $d(P)$.*

Exercise 4.26 *Find two nonisomorphic partial orders whose Hasse diagrams are isomorphic as graphs.*

Exercise 4.27 *How do you form the Hasse diagram of $d(P)$ from that of P?*

Exercise 4.28 *For the topology on $\{a, b, c, d, e\}$ with open sets $\emptyset, \{b\}, \{c\}, \{a, c\}, \{b, d\}, \{a, b, c, d, e\}$, draw the generalized Hasse of the associated preorder.*

Exercise 4.29 *For the preorder whose Hasse diagram is shown below, list the open sets of the associated topology.*

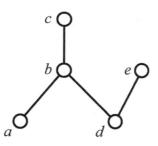

Exercise 4.30 *Prove that given a directed graph D on $[n]$ that is reflexive and transitive, the sets $\mathcal{O}_x = \{y : (y, x)$ is an arc of $D\}$ form the basis of a topology on $[n]$.*

Exercise 4.31 *Prove that the open sets of a topology τ correspond to ideals of $D(\tau)$.*

Exercise 4.32 *A topology τ is $\mathcal{T}_0$ if for every two distinct vertices x and y, there is either an open set containing x and not y, or an open set containing y and not x. Prove that the $\mathcal{T}_0$ topologies on $[n]$ are in a 1-1 correspondence with partial orders on $[n]$.*

Exercise 4.33 *Let τ_1 and τ_2 be topologies on sets V_1 and V_2, respectively. A function $f : V_1 \to V_2$ is **continuous** (from τ_1 to τ_2) if for every open set U of τ_2, $f^{-1}(U)$ is open in τ_1. For preorder $P_1 = (V_1, R_1)$ and $P_1 = (V_2, R_2)$, a function $g : V_1 \to V_2$ is **order preserving** if for x and y in V_1, $(x, y) \in R_1$ implies $(g(x), g(y)) \in R_2$. Prove that $f : V_1 \to V_2$ is continuous (as a map from τ_1 to τ_2) if and only if it is order preserving (as a map from P_1 to P_2).*

Exercise 4.34 *A topology τ on set V is **connected** if and only if V cannot be written as the union of two nonempty open sets. Prove that τ is connected if and only if $D(\tau)$ is weakly connected.*

Exercise 4.35 *Prove Observations 4.2, 4.3 and 4.5.*

Exercise 4.36 *Show that there are two non-homeomorphic topologies τ and σ on $\mathbb{R}$ for which $D(\tau)$ and $D(\sigma)$ are isomorphic as directed graphs.*

Exercise 4.37 *Let P be a preorder on set V and $\mathcal{I}$ its ideals. Show that $\mathcal{I}$ forms a topology on V (such a topology is called an **Alexandroff topology**).*

Exercise 4.38 *Prove that if p is prime, then any topology with exactly p open sets is connected.*

Exercise 4.39 *Prove that the lattice $(Top_V, \subseteq)$ is complete and bounded.*

Exercise 4.40 *Prove that the lattice $\mathcal{T}_{[n]}$ is isomorphic to the* dual *of $Pre_{[n]}$.*

Exercise 4.41 *Prove Theorem 4.8.*

Exercise 4.42 *Prove that almost every partial order has each element in the middle covered above and below.*

Exercise 4.43 *Find a lattice that is bounded but not complemented.*

Chapter 5

Hypergraphs

All of the discrete models we have seen so far – digraphs, graphs, preorders, partial orders – have involved only binary relations. The next step up is to **hypergraphs** (sometimes called **set systems**), which are discrete structures on a set V where the edges are subsets of the power set of V. Hypergraphs originally received attention for their role as models. **Designs** are arrangements of objects subject to particular regularity constraints. **Point set topology** is essentially the study of hypergraphs closed under finite intersection and arbitrary union. **Finite geometry** examines hypergraphs that satisfy a variety of axioms related to Euclidean geometry. And we shall study one of the most important classes of hypergraphs, namely **simplicial complexes** in the next chapter.

5.1 Applying Hypergraphs

We'll begin by examining how hypergraphs have been used as models both inside and outside combinatorics. There are many examples, but we'll restrict ourselves mostly to those involving colourings. (Subsequently, we assume that each edge of our hypergraph has size at least 2.)

5.1.1 Hypergraphs and Graph Colourings

We have already seen Erdös' theorem on the existence of graphs with high chromatic number and high girth. Yet his proof, on one level, is unsatisfying, in that it tells us that there are many such graphs, without providing a single example! Such is the nature of probabilistic arguments. A young Hungarian student (at the time), Laszlo Lovász, was the first to show how to construct such graphs [60], but his argument relied on extending the problem to hypergraphs, proving the existence, for all r, of r–uniform hypergraph of arbitrarily high girth and chromatic number, and then pulling the result back down to graphs. While his argument is beyond the scope of this text, we shall talk about hypergraph colourings and show how viewing this general setting can provide both broad results as well as new constructions for graph colourings.

A **k–colouring** of a hypergraph H on vertex set V is a map $\pi : V \to \{1, \ldots, k\}$ such that no edge e of H is **monochromatic** under π, that is, $|\pi(e)| \geq 2$ for all edges e of H; for a graph (2–regular hypergraph), this definition coincides with the definition of a graph colouring. Again, the definitions of chromatic number and criticality extend in the natural way up to hypergraphs.

Example: A 3–regular hypergraph F shown in Figure 5.1 (also known as the **Fano plane**, where the edges are the straight lines together with the circle). To see that the Fano plane cannot be coloured with 2 colours, note that if there were a 2–colouring of it, then the three corners of the triangle cannot be coloured the same (as otherwise, the three points on the circle must be coloured with the opposite colour, giving a monochromatic edge), so two points of the triangle are coloured one colour, say colour 1, and the other point is coloured with the other colour, 2. The other point on the edge containing the two points of the triangle coloured 1 must be coloured 2, and hence the middle point must be coloured 1. This forces the two remaining points on the circle to be coloured 2, yielding a monochromatic edge (being the circle). Hence F cannot be coloured with 2 colours. We leave it to the reader to verify that there is a 2-colouring of $F - v$ for any vertex v of F, so that F is vertex 3-critical hypergraph. $\triangle$

One of the uses of hypergraphs is to build graphs with certain colouring properties. For example, it has been of great interest to build vertex k–critical

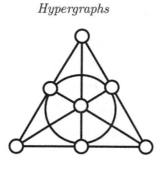

FIGURE 5.1: The Fano plane (edges are indicated by lines or ovals)

graphs for $k \geq 4$, as there is no (and unlikely to be a) good characterization of such graphs. Here is a way to build such graphs from critical hypergraphs and particular graphs called **amenable**. A graph $G = (V, E)$ is **k–amenable** if $\chi(G) = k$ and for any set $S \subseteq \mathbb{N}$ of cardinality k and any nontrivial **k– restraint** function $r : V \to S \cup \{\infty\}$ (that is, r is non-constant except possibly $r(V) = \{\infty\}$) there is a k–colouring $\pi : V \to \{1, \dots, k\}$ such that $\pi(v) \neq r(v)$ for all $v \in V$ (such a colouring is said to be **permitted** by restraint r).

Example: We show that K_n is n–amenable. The result is true for $n = 1$ as the only nontrivial 1 restraint on K_1 assigns ∞ to the only vertex, and it can therefore be coloured with the one colour available. So assume $n \geq 2$ and K_{n-1} is $(n-1)$–amenable. Suppose we have a nontrivial n–restraint $r : V(K_n) \to [n]$. If $r(V(K_n)) = \{\infty\}$, then any n-colouring of K_n is permitted by r. Otherwise, as r is nontrivial, there are vertices u and v with $r(u) \neq r(v)$. Without loss, $r(u) = n$. Then colour v with $n = r(u)$, and consider the $(n-1)$-restraint $r' : V(K_n) - \{v\} \to [n-1]$ where $r'(w) = r(w)$ if $w \neq u$ and $r'(u) = \infty$. Clearly r' is a nontrivial $(n-1)$-restraint on K_{n-1}, so by induction we can extend the colouring to an $(n-1)$-colouring of $V(K_n) - \{v\} \cong K_{n-1}$ permitted by r', and we have an n–colouring of K_n permitted by r.

It follows that K_n is n–amenable. We leave the proof that all odd cycles C_{2n+1} are 3–amenable to the reader (see Exercise 5.4). △

Theorem 5.1 ([96]) *Let H be a k–critical hypergraph on vertex set V ($k \geq$ 3) with edge set F, and for each $e \in F$, let $G_e = (V_e, E_e)$ be a $(k-1)$–amenable*

$(k-1)$–*critical graph of order at least* $|e|$ *(we assume that the* G_E *and* H *are all pairwise disjoint). Form a graph on* $V \cup (\bigcup_{e \in F} V_e)$ *by removing the edges of* H, *keeping the edges of each* G_e, *and for each edge* e *of* H, *joining each vertex of* G_e *to exactly one vertex of* e *such that every vertex of* e *is joined to at least one vertex of* G_e. *Then the resulting graph* G' *is* k–*critical.*

Proof: First we show that G' is k–chromatic. If G' had a $(k-1)$–colouring, then as H is k–chromatic, there must be an edge e of H that is monochromatic, say with colour $k-1$. But then $k-1$ cannot be used on any vertex of G_e as every vertex of G_e is joined to a vertex of e. Thus we must colour G_e with the $k-2$ colours left, a contradiction as $\chi(G_e) = k-1$. Thus $\chi(G') > k-1$. On the other hand, take any k–colouring π of H and let e be an edge of H. Let r be the non-constant k–restraint on the vertices of G_e defined by $r(v) = \pi(u)$ if $u \in e$ and u is adjacent to v in G_e. It is not hard to see that there is a k-colouring of G_e satisfying r. The union of all such colourings, together with π, yields a colouring of G', so that $\chi(G') \leq k$. We conclude that G' is k–chromatic.

We leave it to the reader to determine that G' is connected. It remains to show that $G' - f$ can be $(k-1)$–coloured for any edge f of G'. There are two cases: f is an edge in some G_e, or f is an edge of the form $\{u, v\}$ where u is in some G_e and v is in H. Note that the edge criticality of H ensures that there is a $(k-1)$–colouring π of H such that only edge e is monochromatic, say with colour $k-1$.

For $e' \neq e$, pick a colour $k_{e'}$ that is used by π on some vertex of f. Then the restraint $r_{e'}$ on $G_{e'}$ such that $r_{e'}(v)$ is the colour used by π on the unique neighbour of v in e' if $\pi(v) \in S_{e'} = \{1, \ldots, k\} - \{k_{e'}\}$, and ∞ otherwise. $r_{e'}$ is a non-trivial $(k-1)$-restraint on $G_{e'}$, so there is a colouring of $G_{e'}$ permitted by $r_{e'}$ with the colours in $S_{e'}$.

Such a colouring induces (as above) a nontrivial $(k-)1$–restraint on each $G_{e'}$ for $e' \neq e$, and hence we can extend π to a $(k-1)$–colouring of each $G_{e'}$, $e' \neq e$. In the first case, where f belongs to G_e, we use the criticality of G_e to $(k-2)$–colour it, completing the $(k-1)$–colouring of G'. In the second case, we use the fact that criticality implies that there is a $(k-1)$–colouring of G_e

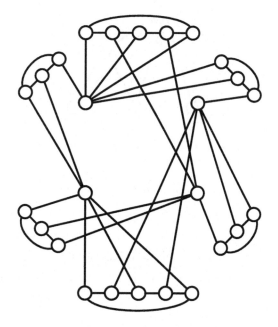

FIGURE 5.2: A 4–critical graph formed via amenable graphs

such that only vertex u gets colour $k - 1$; this yields again a $(k - 1)$–colouring of G'. Thus $\chi(G') \leq k - 1$ and we conclude that G' is k–critical. ■

Starting with the 4–critical (hyper)graph K_4, a 4–critical graph constructed via Theorem 5.1 is shown in Figure 5.2.

5.1.2 Hypergraphs and Generalized Ramsey Theory

We turn to another example. In **generalized Ramsey theory** one wishes to colour objects in a graph with one of k colours so that no matter how this is done, some subobject is monochromatic (this is opposite of what we are trying to do in chromatic theory). Folkman [43] proved that for any positive integer k and any graph F, there is another graph G such that for any function $f : V(G) \to \{1, \ldots, k\}$, some induced copy of F is monochromatic (we call such a graph a **k–Folkman graph** for F, and write $G \to_v^k F$; the subscript "v" denotes that it is the vertices we are colouring).

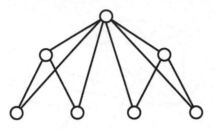

FIGURE 5.3: A 2–Folkman graph for P_3

Example: Consider the graph G shown in Figure 5.1.2. For any function f : $V(G) \to \{1, 2\}$, there is a monochromatic induced copy of P_3. If the top vertex is coloured without loss with 1, then either both P_3's contain a vertex coloured 1, in which case there is a monochromatic induced copy of P_3 coloured 1, or one of the P_3's is monochromatically coloured 2. Thus we see that $G \to_v^2 P_3$, and G is a 2–Folkman graph for P_3. △

Here is an example, due to Nĕsĕtril and Rödl [69] of how one can use hypergraphs and their colourings to prove this result. Let F have order n. For a given $k \geq 1$, take a $(k+1)$–chromatic hypergraph H n–uniform with girth at least 3 (Lovász's construction provides for this graph). As H does not have any 2 cycles, there is at most one edge of H through any two vertices of H. Now for each edge e of H place a copy of F; the previous remark ensures that this process is well defined (i.e. we define whether u and v form an edge only in at most one copy). Then each edge of H induces in this new graph G' a copy of F: for any function $f : V(G') \to \{1, \ldots, k\}$ there is some edge e of H that is monochromatic, and thus the copy of F on e is monochromatic.

5.1.3 Designs and Graphs

A **balanced incomplete block design (BIBD)** is a k–uniform hypergraph $\mathcal{D} = (V, E)$ of order v $(k < v)$ and size $b = |E|$ such that

- every vertex belongs to exactly r of the edges (or "blocks"),

- every two vertices are together in exactly λ of the edges, and

- not every k subset of V is an edge.

Such a hypergraph is called a $(\mathbf{v}, \mathbf{b}, \mathbf{r}, \mathbf{k}, \lambda)$–**design**.

Example: The Fano plane (see Figure 5.1) is a $(7, 7, 3, 3, 1)$–design, as it has 7 vertices and blocks, each block consists of 3 vertices, each vertex is in 3 blocks, and every pair of points belong to exactly one block. $\triangle$

The parameters of designs are not independent of one another, as the next result shows (all parameters can be written in terms of v, k and λ).

Proposition 5.2 *Let $\mathcal{D} = (V, E)$ be a (v, b, r, k, λ)–design. Then*

1. $vr = bk$,

2. $r = \lambda(v - 1)/(k - 1)$,

3. $b = \lambda v(v - 1)/(k(k - 1))$.

Proof: Count all ordered pairs (u, e) where u is a vertex on edge e. There are v many choices for u, and then r choices for the edges that contain u, so there are vr many such ordered pairs. On the other hand, there are b choices for e and there are k choices for vertices of e, so that the number of ordered pairs is also bk. Thus

$$vr = bk.$$

Now fix a vertex u. Let's count, again in two different ways, the number of ordered pairs (w, e) where w is another vertex and e is an edge that contains both u and w. On one hand, there are $v - 1$ choices for vertex w, and for each such choice, the pair u and w lie together in λ edges, so there are $(v - 1)\lambda$ many such ordered pairs. On the other hand, there are r edges e that contain u, and each of these have $k - 1$ vertices in them, besides u, so counting this way we see that there are $r(k - 1)$ many such order pairs. Thus

$$(v - 1)\lambda = r(k - 1)$$

and hence

$$r = \lambda(v - 1)/(k - 1).$$

For the last equality, note that from the other two inequalities we must have

$$
\begin{aligned}
\lambda(v-1)/(k-1) &= r \\
&= \frac{bk}{v}
\end{aligned}
$$

which implies that
$$
b = \frac{\lambda v(v-1)}{k(k-1)}.
$$

∎

Note that as all of the parameters v, b, r, k, λ are integers, we must have from Proposition 5.2 that the following two conditions hold for any BIBD:

$$
\begin{aligned}
\lambda(v-1) &\equiv 0 \pmod{k-1} \\
\lambda v(v-1) &\equiv 0 \pmod{k(k-1)}.
\end{aligned}
$$

While these conditions are necessary for a BIBD, are they sufficient? Wilson's Theorem [105] shows that for $\lambda = 1$ they are *asymptotically* sufficient, in that provided v is large enough (compared to k) there exists a $(v, b, r, k, 1)$–design (his result actually proves more, as the result extends to various size edges).

Designs arise in the setting of experiments, where it would be important, for example, to have every pair of a set of drugs tested for pairwise interactions, while minimizing costs by having individuals testing more than two drugs at a time (if there is a side effect, further investigation can be carried out on the drugs tested in the corresponding block).

Designs are highly structured, and can be used to build other highly structured discrete structures. A **strongly regular graph** with parameters r, λ and γ is an r–regular graph that is not a complete or empty graph, for which every pair of adjacent vertices have λ common neighbours, while every pair of nonadjacent vertices have γ adjacent neighbours. For example, C_5 is strongly regular with parameters 2, 0 and 0, and the complete multipartite graph $K_{n,n,\ldots,n}$ with k parts is strongly regular with parameters $n(k-1)$, $n(k-2)$ and $n(k-1)$.

Balanced incomplete block designs can be used to construct strongly regular graphs as follows. If D is a $(v, b, r, k, 1)$–design, then create a graph whose vertices are the blocks of D, with two such vertices adjacent if and only if the corresponding blocks intersect (such a graph is often called the **line graph** of the design). Then it is not hard to see (Exercise 5.11) that the resulting graph is strongly regular, with parameters

$$\frac{k(v-k)}{k-1}, \frac{v-2k+1}{k-1} + (k-1)^2, k^2.$$

Example: Any complete graph K_n, whose blocks are the edges of the graph, is a $(n, \binom{n}{2}, n-1, 2, 1)$–design. The line graph of the design is precisely the line graph of K_n, and it is not hard to verify that it is a strongly regular graph with parameters $r = 2n - 2$, $\lambda = n - 2$ and $\gamma = 4$. △

Example: Consider the design on $\mathbb{Z}_3 \times \mathbb{Z}_3$, whose blocks are

$$\{(0,0), (0,1), (0,2)\}, \ \{(1,0), (1,1), (1,2)\}, \ \{(2,1), (2,1), (2,2)\},$$

$$\{(0,0), (1,0), (2,0)\}, \ \{(0,1), (1,1), (2,1)\}, \ \{(0,2), (1,2), (2,2)\},$$

$$\{(0,0), (1,1), (2,2)\}, \ \{(1,0), (2,1), (2,2)\}, \ \{(2,0), (0,1), (1,2)\},$$

$$\{(0,0), (1,2), (2,1)\}, \ \{(1,0), (2,2), (0,1)\}, \ \{(2,0), (0,2), (1,1)\}.$$

This is a $(9, 12, 4, 3, 1)$–design. The line graph of the block design is shown in Figure 5.1.3 and is indeed strongly regular. △

Designs can also be connected to various edge decompositions of graphs. For example, an edge decomposition of the complete graph K_v into triangles is, from another viewpoint, a $(v, \binom{v}{3}, (v-1)/2, 3, 1)$–design. More generally, a $(v, \lambda v(v-1)/(k(k-1)), (v-1)/(k-1), k, \lambda)$–design exists if and only if there is an edge decomposition of K_v^λ (the multigraph formed from K_v by replacing every edge by a bundle of λ edges) into cliques of order k.

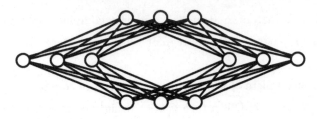

FIGURE 5.4: Line graph of affine plane of order 3

Finally, it remains to find constructions of balanced incomplete block designs, and we delve into linear algebra to construct some arising as finite geometries. A **finite projective plane of order p** is a finite $(p+1)$–uniform hypergraph that satisfies the following three axioms:

1. Any two distinct vertices belong to exactly one edge together.

2. Any two distinct edges intersect in exactly one point.

3. There are four points, no three of which are on an edge together.

It is not hard to verify that the Fano plane once again, serves as an example, as it is a finite projective plane of order 2. In Exercises 5.12 and 5.13 you are asked to verify that any finite projective plane is indeed a BIBD, and moreover, that there is a positive integer n, called the **order**, such that

$$v = b = n^2 + n + 1 \quad \text{and} \quad r = k = n + 1.$$

When do projective planes of order n exist? This is a difficult problem. Examples can be found, however, of prime power order (c.f. [7]). In fact, it is a major open question whether projective planes of non-prime power order exist.

Theorem 5.3 *For every prime power $q = p^k$ (p prime and $k \geq 1$) there is a projective plane of order p.*

Proof: Let $F = \mathbb{F}_q$ be the finite field on q elements (a unique finite field exists on any prime power number of elements – see, for example, [58]). Set $V = \{(x_1, x_2, x_3) : x_i \in F, i = 1, 2, 3\}$, the set of all triples of elements of F.

V is a 3–dimensional vector space over F. Now we take as our set of points X for our finite projective plane the lines (that is, 1–dimensional subspaces) of V (the line generated by $\mathbf{x} \neq \mathbf{0}$ is $F\mathbf{x} = \{f\mathbf{x} : f \in F\}$), and our blocks are the 2–dimensional subspaces of V.

It is not hard to check that for any two distinct points $F\mathbf{x}$ and $F\mathbf{y}$ of X there is a unique block containing them, namely the two dimensional space spanned by $\mathbf{x}$ and $\mathbf{y}$ (these vectors are linearly independent as otherwise one is a scalar multiple of the other and hence the two lines are identical). Furthermore (see Exercise 5.14) any two distinct blocks intersect in exactly one point. Finally, if $\mathbf{v}_1, \mathbf{v}_2, \mathbf{v}_3$ is any vector space basis for V, then the four points $F\mathbf{v}_1$, $F\mathbf{v}_2$, $F\mathbf{v}_3$ and $F(\mathbf{v}_1 + \mathbf{v}_2 + \mathbf{v}_3)$ are distinct and no three are in the same block (see Exercise 5.15).

As all of the axioms are satisfied, it follows that we have a finite projective plane. The number of points is $\frac{q^3-1}{q-1} = q^2 + q + 1$, as each line contains $q - 1$ nonzero vectors and each of the $q^3 - 1$ nonzero vectors is on exactly one line. It follows that the plane has order q (see Exercise 5.16). ∎

We'll return to projective planes in Section 5.2.3, but for now we'll look at another application of hypergraphs.

5.1.4 Hypergraphs and Dimension of Partial Orders

Recall from the previous chapter (Exercise 4.16) that the dimension of a finite partial order P is the minimum number of linear extensions whose intersection is P. For example, the dimension of the partial order shown on the left in Figure 5.5 is 2, as it is not a linear order, but it is the intersection of the two linear orders on the right.

Equivalently, from Exercise 4.18, the dimension of a finite partial order P is the minimum number of linear orders that reverses the critical pairs of P. On the other hand, all critical pairs are pairs of incomparable elements, and we have also seen that a set of such pairs extends to a partial order (and hence to a linear order) if and only if it contains no strict alternating cycle (see Exercises 4.24 and 4.13). Thus let's form a hypergraph H_P^{Crit} on the vertex

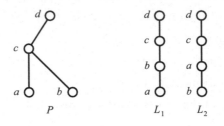

FIGURE 5.5: A 2–dimensional partial order P, as $P = L_1 \cap L_2$

set $\mathrm{Crit}(P)$, the critical pairs of P, with $S \subseteq \mathrm{Crit}(P)$ an edge if and only if the dual of S (formed by reversing all of the ordered pairs in S) is a strict alternating cycle of P. Then the sets of critical pairs not containing a strict alternating cycle are in a 1–1 correspondence with the independent sets of H_P^{Crit}, so the following result follows.

Theorem 5.4 *The dimension of a finite partial order $P = (V, \preceq)$ is equal to the chromatic number of the hypergraph H_P^{Crit}.*   ∎

This modeling of partial order dimension by chromatic number of a hypergraph is very useful. For example, let's show that determining whether a partial order has dimension at most 2 can be carried out in polynomial time (this result is due to Golumbic [50], though the first part of the argument is due to [35] and [49] – see also [97]).

Theorem 5.5 *For a finite partial order $P = (V, \preceq)$, determining whether $dim(P) \leq 2$ can be carried out in polynomial time.*

Proof: We assume in the argument below that P is a partial order that is not a linear order (P is a linear order if and only if its dimension is 1, a trivial case).

For any hypergraph H on vertex set V, we can restrict H down to a graph $\mathrm{Gr}(H)$ on V whose edges are the edges of H of cardinality 2. As the edges of $\mathrm{Gr}(H)$ are a subset of the edges of H, any colouring of H is automatically a colouring of $\mathrm{Gr}(H)$, so we conclude that $\chi(\mathrm{Gr}(H)) \leq \chi(H)$. We shall show that $dim(P) \leq 2$ if and only if $\mathrm{Gr}(H_P^{\mathrm{Crit}})$ is 2–colourable; this will suffice as the graph $\mathrm{Gr}(H_P^{\mathrm{Crit}})$ (with at most $\binom{|V(P)|}{2}$ many vertices) can easily be constructed in polynomial time, and from the chapter on graphs, we know that 2–colourability is polynomial.

Now one direction is clear – if $\dim(P) \leq 2$, by Theorem 5.4 H_P^{Crit} is 2–colourable, so the spanning subhypergraph $\mathrm{Gr}(H_P^{\mathrm{Crit}})$ is also 2–colourable. We now assume that $\mathrm{Gr}(H_P^{\mathrm{Crit}})$ has a 2–colouring $\pi : V(\mathrm{Gr}(H_P^{\mathrm{Crit}})) = V(H_P^{\mathrm{Crit}}) \to \{1,2\}$; by Theorem 5.4 it suffices to show that π is in fact a 2–colouring of H_P^{Crit}. If π is not a 2–colouring of H_P^{Crit}, then there is a monochromatic edge; choose one e_1 of minimum cardinality k (of course, $k \geq 3$ as π properly colours the edges of cardinality 2).

Here is the key observation. For any monochromatic edge e of minimum cardinality k, write $e = \{(x_0, y_0), \ldots, (x_{k-1}, y_{k-1})\}$ where (y_0, x_0), $\ldots, (y_{k-1}, x_{k-1})$ is a strict alternating cycle in P. For any $l \neq j+1$, x_j and y_l are incomparable (Exercise 4.11), so by Exercise 4.17, there is a critical pair (u, v) with $u \preceq x_j$ and $y_l \preceq v$. Now $(u,v), (x_{j+1}, y_{j+1}), \ldots, (x_{l-1}, y_{l-1})$ is the dual of an alternating cycle, and has cardinality less than k. It follows by the minimality of e that $\{(u,v), (x_{j+1}, y_{j+1}), \ldots, (x_{l-1}, y_{l-1})\}$ can't be monochromatic, so as all of $(x_{j+1}, y_{j+1}), \ldots, (x_{l-1}, y_{l-1})$ are coloured $\pi(e)$, such a (u,v) must be coloured the opposite colour, which is $3 - \pi(e)$.

Let's write $e_1 = \{(x_0, y_0), \ldots, (x_{k-1}, y_{k-1})\}$, where $(y_0, x_0), \ldots, (y_{k-1}, x_{k-1})$ is a strict alternating cycle in P. Now for each $l = 0, \ldots, k-1$, choose a critical pair (u_l, v_l) with $y_l \preceq v_l$ and $u_l \preceq x_{l+1}$. Then you can verify that $(u_0, v_0), \ldots, (u_{k-1}, v_{k-1})$ is the dual of an alternating cycle, and moreover, all of the ordered pairs are coloured $3 - \pi(e_1)$, from the key observation. It follows by minimality of k that it is a strict alternating cycle and that k is therefore odd. Let $e_2 = \{(u_0, v_0), \ldots, (u_{k-1}, v_{k-1})\}$; then f is a monochromatic edge of H_P^{Crit} of colour $\pi(e_2) = 3 - \pi(e_1)$.

Finally, choose a critical pair (s,t) with $s \preceq u_{k-1}$ and $v_3 \preceq t$. From the key observation, using the monochromatic edge e_2, coloured 2, and critical pair (s,t), we see that (s,t) must be coloured $3 - \pi(e_2) = \pi(e_1)$. On the other hand, $s \preceq u_{k-1} \preceq x_0$ and $y_3 \preceq v_3 \preceq t$. Using monochromatic edge e_1, coloured $\pi(e_1)$, and critical pair (s,t), we see that (s,t) must be coloured $3 - \pi(e_1)$.

This contradiction implies that indeed π is a 2–colouring of H_P^{Crit}, and so $\chi(H_P^{\text{Crit}}) \leq 2$. It follows by Theorem 5.4 that $\dim(P) \leq 2$. ∎

5.2 Modeling Hypergraphs

We'll turn the tables now and look at ways to model hypergraphs. In the next chapter, we'll see other models of particular classes of hypergraphs (simplicial complexes), but here we'll investigate in particular how different algebraic models can help prove theorems on hypergraphs; our emphasis is on the variety of algebraic models and techniques available.

5.2.1 Criticality and Matrix Rank

We begin by proving a result on the number of edges in a 3–critical r–uniform hypergraph. Of course, the only interesting case is $r \geq 3$ as for $r = 2$, the 3–critical graphs, namely the odd cycles, are known completely (r–uniform hypergraph 2–colourability is **NPc** for $r \geq 3$ [61], and thus there is unlikely to be a nice characterization of 3–critical hypergraphs). You can check that the Fano plane is indeed a 3–uniform 3–critical hypergraph.

How many edges can a 3–critical r–uniform hypergraph of order n have?

Theorem 5.6 ([65]) *If H is a 3–critical r–uniform hypergraph of order n with m edges, then $m \leq \binom{n}{r-1}$.*

Proof: Let the vertices of H be V, the edges be $e_1, \ldots, e_m$, and let $S_1, \ldots, S_{\binom{n}{r-1}}$ be a listing of all the $r-1$–subsets of V. Form the $\binom{n}{r-1} \times m$ matrix $S = (s_{i,j})$ where

$$s_{i,j} = \begin{cases} 1 & \text{if } S_i \subset E_j \\ 0 & \text{otherwise.} \end{cases}$$

We claim that S has **full column rank**, i.e. the columns $\mathbf{e_1}, \ldots, \mathbf{e_m}$ of S are linearly independent (over $\mathbb{R}$). This will complete the argument as then the row rank, which equals the column rank, must be m as well, so that the number of rows, namely $\binom{n}{r-1}$, must be at least m.

Suppose that the columns of S are not linearly independent. Then there exist $\alpha_1, \ldots, \alpha_m$, not all zero, such that

$$\sum_{i=1}^{m} \alpha_i \mathbf{e_i} = \mathbf{0};$$

without loss, we assume that $\alpha_1 \neq 0$. It follows that for all $j = 1, \ldots, \binom{n}{r-1}$, we have

$$\sum_{e_i \supset S_j} \alpha_i = 0. \tag{5.1}$$

Take a 2–colouring of $H - e_1$; in such a colouring, e_1 is the only edge of H that is monochromatic. Thus we can partition V into two sets V_1 and V_2 such that $e_1 \subseteq V_1$ and no other edge of H is contained in V_1 or V_2. Let

$$I_\nu = \{i : |e_i \cap V_1| = \nu\},$$

$$J_\nu = \{j : |S_j \cap V_1| = \nu\}$$

and

$$\beta_\nu = \sum_{i \in I_\nu} \alpha_i.$$

From (5.1) we see that for any ν,

$$\sum_{j \in J_\nu} \sum_{e_i \supset S_j} \alpha_i = 0.$$

What edges contribute to this sum? Well, any $r - 1$–subset S_j whose intersection with V_1 has cardinality ν that is contained in an edge (of cardinality r) of H implies that such edge must intersect V_1 in either ν vertices or $\nu + 1$ vertices. Thus if we take an edge e_i that intersects V_1 in ν vertices, it contains $r - \nu$ many such S_js (as we simply drop out a vertex of e_i that is in V_2 in that many ways). Similarly, if we take an edge e_i that intersects V_1 in $\nu + 1$ vertices, it contains $\nu + 1$ many such S_js (as we simply drop out a vertex of e_i that is in V_1 in that many ways). Thus we deduce that

$$(r - \nu) \sum_{i \in I_\nu} \alpha_i + (\nu + 1) \sum_{i \in I_{\nu+1}} \alpha_i = 0,$$

that is,

$$(r - \nu)\beta_\nu + (\nu + 1)\beta_{\nu+1} = 0.$$

Therefore, we have

$$\beta_{\nu+1} = -\frac{r - \nu}{\nu + 1}\beta_\nu$$

which implies that

$$\beta_\nu = \left(-\frac{r - (\nu - 1)}{\nu}\right) \cdots \left(-\frac{r - 1}{2}\right)\left(-\frac{r - 0}{1}\right)\beta_0$$
$$= (-1)^\nu \binom{r}{\nu}\beta_0.$$

In particular,

$$\beta_r = (-1)^r \beta_0,$$

so that

$$\sum_{e_i \subseteq V_1} \alpha_i = (-1)^r \sum_{e_i \subseteq V_2} \alpha_i.$$

However, the right side is clearly 0 as the sum is empty, while the left side is α_1, as the only edge e contained in V_1 is e_1. But this contradicts $\alpha_1 \neq 0$.

We conclude that indeed S has rank m, so we are done.   ∎

5.2.2   Criticality and Multilinear Algebra

Recall that in section 3.2.2 we presented Lovász's argument that a τ–critical graph G with $\tau(G) = t$, has size at most $\binom{t+1}{2}$; the argument was via matrices. Bollobás extended the notion of τ–critical graph to hypergraphs in the following way. A **transversal** of a hypergraph $H = (V, E)$ is a subset T of V such that for all edges e of H, $T \cap e \neq \emptyset$; the minimum cardinality of a transversal of H is written as $\tau(H)$. A hypergraph is τ–**critical** if $\tau(H') < \tau(H)$ for every subhypergraph H' of H. Bollobás [13] proved the following:

Theorem 5.7 *If* $H = (V, E)$ *is an* r–*uniform* τ–*critical hypergraph with* $\tau(H) = t$, *then*

$$|E| \leq \binom{r + t - 1}{r}.$$

We present a proof that is due to Lovász [62] – a lovely and surprising stroll through multilinear algebra! We'll need some background first. Suppose that W is a finite–dimensional vector space of dimension d. Then for any $k \leq d$, we can form a new vector space, called $\mathbf{k^{th}}$ **exterior (or alternating) power** of W, as a quotient of the tensor product of W with itself k times:

$$\wedge^k V = \otimes_{i=0}^k W / \{x_1 \otimes \ldots \otimes x_k : x_i = x_j \text{ for some } i \neq j\}.$$

If you haven't seen exterior products (or even tensor products before), the important facts about the vector space $\wedge^k W$ (see, for example, [58, Chapter 13]) are:

- If $\{\mathbf{v}_1, \ldots, \mathbf{v}_d\}$ is a basis for W, then

$$\{\mathbf{v}_{i_1} \wedge \ldots \wedge \mathbf{v}_{i_k} : 1 \leq i_1 < i_2 < \ldots < i_k \leq d\}$$

  is a basis for $\wedge^k W$, so that $\dim \left(\wedge^k W \right) = \binom{d}{k}$.

- $\mathbf{x}_1 \wedge \ldots \wedge \mathbf{x}_k = 0$ if and only if $\{\mathbf{x}_1, \ldots, \mathbf{x}_k\}$ is linearly dependent in W.

We now return back to Lovász' proof of Bollobás' Theorem. Given an r–uniform τ–critical hypergraph $H = (V, E)$ with $\tau(H) = t$ and $n = |V|$, let's take $W = \mathbb{R}^{r+t-1}$ and a set of vectors $\{\mathbf{v}_1, \ldots, \mathbf{v}_n\}$ in general position, that is, every subset of $r + t - 1$ vectors is linearly independent. For a subset S of V, we represent S as a exterior power:

$$\bigwedge S = \wedge_{v_i \in S} \mathbf{v}_i.$$

You'll see shortly how well exterior products can be used to model intersection and non-intersection of sets.

For any edge e of H, by criticality there is a transversal T_e of size $t - 1$ of $H - e$; note that in H, T_e meets every edge of H but e. We claim now that the set $\{\bigwedge T_e : e \in E\}$ is linearly independent in $\wedge^{r+t-1} W$. For suppose that for some $\alpha_e \in \mathbb{R}$, we have

$$\sum_{e \in E} \alpha_e (\bigwedge T_e) = \mathbf{0}.$$

Then for any edge f we have

$$\left(\bigwedge f\right) \wedge \left(\sum_{e \in E} \alpha_e \left(\bigwedge T_e\right)\right) \; = \; \bigwedge f \wedge \mathbf{0},$$

that is,

$$\left(\sum_{e \in E} \alpha_e \left(\bigwedge f\right) \wedge \left(\bigwedge T_e\right)\right) \; = \; \mathbf{0}. \tag{5.2}$$

However, if $e \neq f$ then T_e intersects f and thus the multiset $\{\mathbf{v} : v \in f \cup T_e\}$ is linearly dependent; if, on the other hand, if $e = f$, then T_e and f are disjoint, and hence by the general position of the vectors $\{\mathbf{v}_1, \ldots, \mathbf{v}_n\}$, we have that $\{\mathbf{v} : v \in f \cup T_e\}$ is linearly independent. It follows from (5.2) and the properties of the exterior product that

$$\alpha_e \left(\bigwedge e\right) \wedge \left(\bigwedge T_e\right) = \mathbf{0},$$

which implies that $\alpha_e = 0$, as $\left(\bigwedge e\right) \wedge \left(\bigwedge T_e\right) \neq \mathbf{0}$. Thus all $\alpha_e = 0$, implying that the set $\{\bigwedge T_e : e \in E\}$ is linearly independent in $\bigwedge^{r+t-1} W$. This completes the proof as the cardinality of this set can be no more than the dimension of W, that is, $|E| \leq \binom{r+t-1}{r}$. $\blacksquare$

5.2.3 Finite Geometries and Orthogonality

In [22] projective planes are used to construct Folkman graphs. In one part of the argument, there is a need to estimate, for a given large subset X of points, the sum of the number of points that each line in a somewhat large collection of lines intersect X in. The argument relies on forming a specific matrix whose entries were chosen to make the rows orthogonal. Recall that from Theorem 5.3, for any prime p, we can find a projective plane $\mathcal{P} = (V, L)$ of order p.

Proposition 5.8 ([22]) *Let be $\mathcal{P} = (V, L)$ a projective plane of prime order p, with $N = p^2 + p + 1$ points and lines. Let $\alpha \in (0, 1)$ and let X be a subset of points of cardinality $m = \lfloor \alpha N \rfloor$. Let $\mathcal{L}$ be a subcollection of lines of L of* *cardinality $L \geq N^{1/2+\varepsilon}$ for some $\varepsilon \geq 0$ (without loss, $\mathcal{L} = \{l_1, \ldots, l_L\}$). Then*

$$\sum_{l \in \mathcal{L}} |l \cap X| \sim \alpha \sqrt{N} L.$$

Proof: Let the points and lines of $\mathcal{P}$ be $p_1, \ldots, p_N$ and $l_1, \ldots, l_N$, respectively, with $X = \{p_1, \ldots, p_m\}$. For $l \in L$, define

$$x_l = \sum_{l \in \mathcal{L}} |l \cap X|.$$

We form an $N \times N$ matrix B whose rows and columns are indexed by the points and lines of $\mathcal{P}$ with

$$B_{i,j} = \begin{cases} \lambda & \text{if } p_i \in l_j \\ -1 & \text{otherwise;} \end{cases}$$

we shall see how to choose a useful λ shortly.

Let $\mathbf{b}_i$ denote the ith row of B. For $i \neq j$ from the properties of projective planes we deduce that

$$\begin{aligned} \mathbf{b}_i \cdot \mathbf{b}_j &= \lambda^2 + 2p\lambda(-1) + (p^2 + p + 1 - 2p - 1)(-1)^2 \qquad (5.3) \\ &= \lambda^2 - 2p\lambda + p^2 - p. \end{aligned}$$

We now choose λ so that the vectors $\{\mathbf{b}_i : i = 1, \ldots, N\}$ form an orthogonal set; we'll choose

$$\lambda = p + \sqrt{p}$$

(though $\lambda = p - \sqrt{p}$ would also do).

By orthogonality and (5.4) we have

$$\begin{aligned} \| \mathbf{b}_1 + \ldots + \mathbf{b}_m \|^2 &= \| \mathbf{b}_1 \|^2 + \ldots + \| \mathbf{b}_m \|^2 \\ &= m((p+1)\lambda^2 + p^2) \\ &= m((p+1)(p^2 + 2\sqrt{p} + p) + p^2) \\ &\sim \alpha N^{5/2}. \end{aligned}$$

On the other hand, let $\mathbf{c}_i$ be the restriction of $\mathbf{b}_i$ to the columns of $\mathcal{L}$, and set

$$\mathbf{c}_1 + \ldots + \mathbf{c}_m = (\psi_1, \ldots, \psi_L),$$

so that

$$\psi_l = \lambda x_l - (m - x_l).$$

A calculation shows that

$$\| \mathbf{c}_1 + \ldots + \mathbf{c}_m \|^2 \;=\; \sum_{l \in \mathcal{L}} \psi_l^2 \qquad (5.4)$$

$$\geq \; \frac{1}{L} \left(\sum_{i=1}^{l} \psi_i \right)^2 \qquad (5.5)$$

$$= \; \frac{1}{L} \left((\lambda + 1) \sum_{l \in \mathcal{L}} x_l - mL \right)^2$$

$$= \; \left(\frac{\lambda + 1}{\sqrt{L}} \sum_{l \in \mathcal{L}} x_l - m\sqrt{L} \right)^2 . \qquad (5.6)$$

Clearly

$$\| \mathbf{b}_1 + \ldots + \mathbf{b}_m \|^2 \geq \| \mathbf{c}_1 + \ldots + \mathbf{c}_m \|^2$$

so it follows from (5.4) and (5.6) that

$$\left| \frac{\lambda + 1}{\sqrt{L}} \sum_{l \in \mathcal{L}} x_{l \in calL} - m\sqrt{L} \right| \leq \sqrt{m((p+1)(p^2 + 2\sqrt{p} + p) + p^2)} \sim \sqrt{\alpha} N^{5/4}.$$

Now $m\sqrt{L} \sim \alpha N^{5/4+\varepsilon/2}$ which is much bigger than $\sqrt{\alpha} N^{5/4}$, so we must have

$$\frac{\lambda + 1}{\sqrt{L}} \sum_{l \in \mathcal{L}} x_l \sim m\sqrt{L},$$

that is,

$$(\lambda + 1) \sum_{l \in \mathcal{L}} x_l \sim mL = \alpha NL.$$

As $\lambda + 1 \sim \sqrt{N}$, we finally have

$$\sum_{l \in \mathcal{L}} x_l \sim \alpha \sqrt{N} L.$$

■

5.2.4   Designs from Codes

Building designs can be hard work! There are various constructions for building designs of different kinds. We will describe here one method for building one class of designs. A **Steiner** (t, k, n)**- system** is a k–uniform hypergraph $\mathcal{D} = (V, E)$ of order n $(k < v)$ such that every t-subset of V is contained in exactly one edge. For example, the Fano plane discussed in Section 5.1.3 is a Steiner $(2, 3, 7)$-system (any (v, b, r, k, λ)–design is by definition a Steiner $(2, k, v)$-design).

In order to build designs, one often uses algebraic structures, and for Steiner systems, we can use codes to aid us. A **(binary) code** of length n is a collections of vectors (called **codewords**) in $\mathbb{Z}_2^n$ (one can more generally form codes over other fields, but for our purposes here, we restrict ourselves to working over $\mathbb{Z}_2$).

Example: Two codes of length 7 are $\mathcal{C}_1 = \mathbb{Z}_2^7$ and $\mathcal{C}_2 =$

$\{(0000000, 111000, 1001100, 1000011, 0101010, 0011001, 0010110, 0100101,$

$1101001, 1100110, 0001111, 1011010, 0010011, 0111100, 1010101, 1111111\}$.

(In this section, we often write, as is usual for codes, vectors without brackets and commas). The code $\mathcal{C}_2$ is known as the **Hamming** $(7, 4)$**-code**. (The 7 comes from the length of the code, and the 4 from the fact that it is, indeed, a vector space of dimension 4, and hence has $2^4 = 16$ vectors.)   △

The code $\mathcal{C}_1$ clearly has the most codewords that one can possess in $\mathbb{Z}_2^7$, namely $2^7 = 128$, many more than the 16 that code $\mathcal{C}_2$. However, $\mathcal{C}_1$ is very unforgiving when it come to errors. If 0101010 (a codeword under either scheme) is sent but 0101110 is received, there is no way of detecting, in code $\mathcal{C}_1$, whether an error was made in transmission, as 0101110 is a valid codeword in $\mathcal{C}_1$. On the other hand, under $\mathcal{C}_2$ we would recognize, upon receiving 0101110, that a error in transmission must have been made, as 0101110 is not a codeword. Moreover, if we assume or know that the most likely error is just that of a single digit (see Problem 5.21), we could scan through our 16 codewords and realize that the unique closest codeword to the one received was 0101010 (as it is the only codeword of C_2 that differs from the one received in a single digit). We conclude that the message received was in error, and if there was only a single error, we deduce that the message sent was 0101010.

Thus there is a tradeoff between the number of codewords of a given length and the ability to detect and correct errors. An **e-error correcting code** is a code for which any error in the transmission of a codeword that results at most e digits errors can be corrected properly. Much of the work in cryptography has developed around creating error correcting code and quick algorithms for decoding the correct codeword sent.

For two codewords $\mathbf{x}, \mathbf{y} \in \mathbb{Z}_2^n$, we define the **distance** between them, $d(\mathbf{x}, \mathbf{y})$, as the number of coordinates in which they differ. For example, $d(0101010, 0101110) = 1$ as they differ in only the fifth coordinate. It is not hard to verify that the distance function is a **metric** on vectors of length n, namely

1. $d(\mathbf{x}, \mathbf{x}) = 0$,

2. $d(\mathbf{x}, \mathbf{y}) = d(\mathbf{y}, \mathbf{x})$, and

3. $d(\mathbf{x}, \mathbf{y}) + d(\mathbf{y}, \mathbf{z}) \geq d(\mathbf{x}, \mathbf{z})$

for all vectors $\mathbf{x}, \mathbf{y}, \mathbf{z} \in \mathbb{Z}_2^n$ (see Exercise 5.23). We also define the **weight** w of a vector $\mathbf{x} \in \mathbb{Z}_2^n$ as the number of nonzero components of $\mathbf{x}$, so, for example, $w(0101010) = 3$.

For any codeword $\mathbf{c} \in \mathbb{Z}_2^n$ and any nonnegative integer e, we define the **sphere of radius e** centered at $\mathbf{x}$ as

$$S(\mathbf{c}, e) = \{\mathbf{y} \in \mathbb{Z}_2^n : d(x, y) \leq e\}.$$

The key observation is that the minimum distance between codewords in a code C is at least $2e + 1$ iff the code is an e-error correcting code (see Exercise 5.25).

Example: It is laborious but easy to check that the Hamming $(7, 4)$-code

$$\{(0000000, 111000, 1001100, 1000011, 0101010, 0011001, 0010110, 0100101,$$

$1101001, 1100110, 0001111, 1011010, 0010011, 0111100, 1010101, 1111111\}$

described earlier has minimum distance of 3 between codewords. Therefore it is a 1-error correcting code. △

The Hamming $(7, 4)$ code has an even stronger property than just being a 1-error correcting code. An e-error correcting code $\mathcal{C}$ of length n is **perfect** if every vector $\mathbf{x} \in \mathbb{Z}_2^n$ is within e of some codeword, that is, the spheres

$$\{S(\mathbf{c}, e) : \mathbf{c} \in \mathcal{C}\}$$

partition $\mathbb{Z}_2^n$, the set of all vectors of length n. Why is the Hamming $(7, 4)$-code perfect? Well, the 16 spheres $S(\mathbf{c}, 1)$, as c ranges over all 16 codewords, are clearly disjoint as the minimum distance e between codewords is 3, and each such sphere contains precisely $1 + 7 = 8$ vectors. The total number of vectors in all spheres is therefore $16 \times 8 = 128$, the number of vectors in $\mathbb{Z}_2^7$, so every vectors of $\mathbb{Z}_2^7$ lies in precisely one sphere.

We can form a hypergraph from the Hamming $(7, 4)$-code. Let the vertex set of H be $\{1, 2, 3, 4, 5, 6, 7\}$. For every codeword $\mathbf{c}$ of weight 3 (the minimum weight of a nonzero code), take the edge $e_{\mathbf{c}} = \{i : \text{the } i^{th} \text{ coordinate of } \mathbf{c} \text{ is } 1\}$. Thus from our list of codewords we find that the edges of H are

$$\{1, 2, 3\}, \{1, 4, 5\}, \{1, 6, 7\}, \{2, 4, 6\}, \{3, 4, 7\}, \{3, 5, 6\}, \{2, 5, 7\}.$$

A quick check will show that this is a Steiner $(1, 3, 7)$-system.

This connection is part of a more general construction of Steiner systems from codes. To state the theorem, we need the following definition. Given a set of vectors $U \subseteq \mathbb{Z}_2^n$, the **hypergraph supported by U** has vertices $[n] = \{1, 2, \ldots, n\}$ and, for each $\mathbf{u} \in U$, the edge $f = \{i : \text{the } i^{th} \text{ coordinate of } \mathbf{u} \text{ is } 1\}$ (we say as well that $\mathbf{u}$ is the vector supported by f, and f is the edge supported by $\mathbf{u}$).

Theorem 5.9 *Suppose that $\mathcal{C} \subseteq \mathbb{Z}_2^n$ is a perfect e-error correcting code, with $\mathbf{0} \in U$. Let U be the set of codewords of weight $2e + 1$. Then the hypergraph H supported by U is a Steiner $(e + 1, 2e + 1, n)$-system.*

Proof: Clearly H has n vertices and every edge of H has cardinality $2e + 1$. Let t be a subset of $V(H)$ of cardinality $e + 1$; we need to show that t is in a unique edge of H. Let $\mathbf{t}$ be the vector whose support is t; clearly $\mathbf{t}$ has weight $e + 1$.

Now as the code is perfect, there is a codeword $\mathbf{c}_t$ that is at distance at most e from $\mathbf{t}$. Note that $\mathbf{c}_t \neq \mathbf{0}$ as $\mathbf{t}$ has weight $e + 1$, and the weight of $\mathbf{c}_t$ must be at least $2e + 1$ as the code is e-error correcting. As $\mathbf{t}$ has weight $e + 1$ 1's, it follows that $\mathbf{c}_t$ must have weight exactly $2e + 1$.

If $\mathbf{c}_t$ has j 1's in common with $\mathbf{t}$, then $\mathbf{c}_t$ and $\mathbf{t}$ differ in precisely $(e + 1 - j) + (2e + 1 - j) = 3e + 2 - 2j$ coordinates. Thus $3e + 2 - 2j \leq e$ which implies that $2e + 1 \leq 2j - 1$. But as $\mathbf{t}$ has weight $e + 1$, we must have $j \leq e + 1$. If $j \leq e$ then $2e + 1 \leq 2j - 1 \leq 2e - 1$, a contradiction. Thus $j = e + 1$, and hence the edge f supported by $\mathbf{c}_t$ contains t. If there were another such edge f', say supported by codeword $\mathbf{c}'$ of weight $2e + 1$, that contained t as well, then $\mathbf{c}_t$ and $\mathbf{c}'$ would have $e + 1$ 1's in common, and hence these two codewords would have distance $2(2e + 1 - (e + 1)) = 2e$ from each other, a contradiction. It follows that H is a Steiner $(e + 1, 2e + 1, n)$-system. ∎

Thus perfect codes are useful for constructing designs, but Theorem 5.9 is only the tip of the iceberg. The exercise sets will explore more connections between codes and designs. Much more on codes and their connections to designs can be found in [67].

Exercises

Exercise 5.1 *The **complete l-uniform hypergraph** K_n^l of order n is the hypergraph on $[n]$ that contains all l-subsets of $[n]$. What is the chromatic number of K_n^l?*

Exercise 5.2 *We define an independent set of a hypergraph H on vertex set V to be a collection of vertices not containing an edge. Prove that $\chi(H)$ is the minimum number of independent sets that cover V.*

Exercise 5.3 *Show that any k-amenable graph is vertex k-critical.*

Exercise 5.4 *Show that the odd cycles are 3–amenable.*

Exercise 5.5 *Show that G' of Theorem 5.1 is connected.*

Exercise 5.6 *Prove that if G_k is a k–Folkman graph for G, then $G_{k+1} = G[G_k]$ is a $(k+1)$–Folkman graph for G. Conclude that for every graph G of order n and every positive integer k there is a k–Folkman graph for G of order n^k.*

Exercise 5.7 *Show that in the Něsětril–Rödl construction of Section 5.1.2, if H greater than $|F|$ and F is 2–connected (that is, is connected and has no vertex whose removal disconnects F) then the only copies of F are those on the edges of H.*

Exercise 5.8 *Prove that a result due to Seymour [80]: if $k \geq 3$ and H is a k–critical r–uniform hypergraph of order n with m edges, then $m \geq n$. (Hint: Consider the rank of the $n \times m$ vertex–edge matrix of H.)*

Exercise 5.9 *Prove Fisher's Inequality [42] that in any BIBD, $b \geq v$. (Hint: Prove that the vertex–edge matrix A has full row rank by considering the matrix AA^T.)*

Exercise 5.10 *[79] A collection of vectors X of unit length in $\mathbb{R}^d$ is a **two-distance set** if the distance between any two vectors of X is of one of two values. Prove that if X is a two-distance set in $\mathbb{R}^d$, then*

$$|X| \leq \frac{1}{2}d(d+3) + 1.$$

(Hint: Let the distances be α and β. Let $v_1, \ldots, v_d$ be variables, and set $\mathbf{v} = (v_1, \ldots, v_d)$. For each $\mathbf{y} \in X$, define the polynomial

$$p_{\mathbf{y}}(v_1, \ldots, v_d) = \left(\mathbf{y} \cdot \mathbf{v} - \left(1 - \frac{\alpha^2}{2} \right) \right) \left(\mathbf{y} \cdot \mathbf{v} - \left(1 - \frac{\beta^2}{2} \right) \right).$$

Then prove that $\{p_{\mathbf{y}} : y \in X\}$ is linearly independent in the vector space of all real polynomials of degree at most 2 in the d variables $v_1, \ldots, v_d$.)

Exercise 5.11 *Prove that the line graph of a $(v, b, r, k, 1)$–design with $b > v$ is a strongly regular graph with parameters*

$$\frac{k(v - k)}{k - 1}, \frac{v - 2k + 1}{k - 1} + (k - 1)^2, k^2.$$

How many vertices does the graph have?

Exercise 5.12 *Prove that any finite projective plane is a BIBD.*

Exercise 5.13 *Prove that for any finite projective plane there is a positive integer n such that*

$$v = b = n^2 + n + 1 \quad and \quad r = k = n + 1.$$

Exercise 5.14 *Prove why, in the proof of Theorem 5.3, any two distinct blocks intersect in exactly one point.*

Exercise 5.15 *Prove why, in the proof of Theorem 5.3, the four points $F\mathbf{v}_1$, $F\mathbf{v}_2$, $F\mathbf{v}_3$ and $F(\mathbf{v}_1 + \mathbf{v}_2 + \mathbf{v}_3)$ are distinct and no three are in the same block.*

Exercise 5.16 *Prove why, in the proof of Theorem 5.3, the fact that the number of points is $\frac{q^3 - 1}{q - 1} = q^2 + q + 1$ uniquely determines that the plane has order q.*

Exercise 5.17 *Let q be a prime power, and consider the hypergraph on the vector space $\mathbb{Z}_q \times \mathbb{Z}_q$ whose blocks are cosets of lines in this vector space (over $\mathbb{Z}_q$). Show that this hypergraph is a BIBD (it is called an **affine plane** of order q).*

Exercise 5.18 *In the proof of Theorem 5.8, explain how to get from (5.4) to (5.5). (Hint: Think Cauchy–Schwartz!)*

Exercise 5.19 *The exterior product is linear in each component. From this and the definition of $\wedge^k W$, deduce that if $\{\mathbf{x}_1, \ldots, \mathbf{x}_k\}$ is linearly dependent in W, then $\mathbf{x}_1 \wedge \ldots \wedge \mathbf{x}_k = 0$.*

Exercise 5.20 *Suppose that $A_1, \ldots, A_n$ are sets of size a, and $B_1, \ldots, B_n$ are sets of size b, such that $A_i \cap B_j = \emptyset$ if and only if $i = j$. Prove that $n \le \binom{a+b}{b}$, so that the chain cannot be too long! (Note that we have not said anything about the cardinality of the underlying set for the A_i's and B_j's.)*

Exercise 5.21 *Suppose that in the transmission of a single digit, 0 or 1, the probability that the digit is "flipped" (i.e. given a 0 is sent, a 1 is received, or vice versa) is $p \in (0, 1/2)$, and that the correct digit is received $q = 1 - p$. Assume that the probability of a correct digit being received in a certain coordinate is independent of the probabilities in other coordinates. Suppose that $q < p$. If a given vector $\mathbf{x} \in \mathbb{Z}_2^n$ is sent, show that the probability of a specific vector with i errors being received is greater than the probability of a specific vector with $i + 1$ errors being received for $i = 0, 1, \ldots, n - 1$.*

Exercise 5.22 *Under the model for exercise 5.21, find the probability for a 1-error correcting code that the correct code is decoded, and compare it to the probability that the correct code is received.*

Exercise 5.23 *Prove that the distance function is a metric on $\mathbb{Z}_2^n$.*

Exercise 5.24 *If C is a code and $\mathbf{c} \in C$, then the code $C - \mathbf{c}$ consists of all vectors $\mathbf{x} - \mathbf{c}$ where $\mathbf{x}$ is a codeword of C. Prove that $C - \mathbf{c}$ has the same minimum distance between codewords as does C, and that one if perfect iff the other is.*

Exercise 5.25 *Suppose that the minimum distance between codewords of a code C is d. Prove that if d is odd, the code can correct $(d - 1)/2$ errors, and if d is even, it can correct $(d - 2)/2$ errors and detect $d/2$ errors.*

Exercise 5.26 *Prove that the hypergraph supported by the codewords of weight 3 in the Hamming $(7, 4)$-code is isomorphic to the Fano plane.*

Exercise 5.27 *A code $C \subseteq \mathbb{Z}_2^n$ is **linear** if the codewords form a subspace. Prove that the Hamming $(7, 4)$ code is linear, and find a basis for the subspace. What is its dimension? Find also a basis for the orthogonal complement of the subspace.*

Exercise 5.28 *Suppose that $C \subseteq \mathbb{Z}_2^n$ is a code. Form a new code $C' \subseteq \mathbb{Z}_2^{n+1}$, the **extended code** of C, by appending to each codeword $\mathbf{c} \in C$ a new coordinate that is 0 if the sum of the coordinates of $\mathbf{c}$ are even, and 1 otherwise.*

Prove that if C is e-error correcting, then C' is e-error correcting, and moreover, if C has odd minimum distance d between codewords, then the minimum distance between codewords of C' is $d + 1$.

Exercise 5.29 *Construct the extended Hamming $(7, 4)$-code (see Exercise 5.28).*

Chapter 6

Complexes and Multicomplexes

In this final chapter we consider an important subclass of hypergraphs, namely (simplicial) complexes, and their multiset extension, multicomplexes. It is fascinating how often complexes arise in combinatorics, and how useful they are.

Recall that a **(simplicial) complex** $\Delta = (X, E)$ is a hypergraph whose edge set is closed under containment, that is,

$$Y \in E, \ X \subseteq Y \Rightarrow X \in E;$$

thus if E is nonempty, then $\emptyset \in E$. We often assume that the complex $\Delta = (X, E)$ has the property that $\{v\}$ is a face of Δ for all $v \in X$ (this assumption just amounts to removing from X any vertices not in any face).

What is apparent from the definition is that a finite complex is completely determined by its maximal edges; if $\mathcal{B} = \{e \in E : \text{for all } e' \in E, e \subseteq e' \Rightarrow e = e'\}$, then $E = \{X : X \subseteq \mathcal{B}\}$. This is an essential remark, as when we wish to represent a complex for computation, the edge set representation may be exponentially large compared to the more compact maximal set representation.

A **circuit** of $\Delta = (X, E)$ is a "minimal non-face"; that is, $C \subseteq X$ is a circuit if and only if C is not a face of Δ but for every $x \in C$, $C - \{x\}$ is a face of Δ; once again, a complex Δ is completely determined by its circuits, as $E = \{Y \subseteq X : C \not\subseteq Y \text{ for all circuits } C \text{ of } \Delta\}$.

One often calls edges of a complex **faces** or **independent sets**, and maximal faces **facets** or **bases**. A simplicial complex Δ is a **cone** over v if every facet of Δ contains v. Simplicial complexes first arose in the study of topology, where simplicial complexes were used to give a concrete combinatorial construction to topological spaces. Indeed, much of algebraic topology is often developed on simplicial complexes as a stepping stone to topological spaces, and there are some striking application of both homology and homotopy theory on simplicial complexes to combinatorics [64, 63].

For a face σ of complex Δ, we let $\bar{\sigma}$ denote the set of subsets of σ (of course, $\bar{\sigma} \subseteq E(\Delta)$). If σ_1 and σ_2 are any two independent sets of Δ with $\sigma_1 \subseteq \sigma_2$, the **interval** $[\sigma_1, \sigma_2]$ is $\{\sigma : \sigma_1 \subseteq \sigma \subseteq \sigma_2\}$. The **dimension** of a complex Δ is equal to the cardinality of the largest set in Δ (this is one more than the definition of dimension usually found in the algebraic topological literature, but will be convenient for our combinatorial applications). We say that Δ is **purely** d–**dimensional** if Δ is d–dimensional and every maximal independent set (facet or basis) of Δ has cardinality d.

A complex Δ is a **matroid** if the well known **exchange axiom** holds (c.f. [103]):

$$X,\ Y \in E(\Delta), |X| > |Y| \Rightarrow \text{for some } x \in X - Y)(Y \cup \{x\} \in E(\Delta).$$

The matroid axiom arose from extracting one of the fundamental properties of linearly independent sets of vectors, which form simplicial complexes (it is in fact the exchange axiom that plays the key role in the proof that any two bases of a finite dimensional vector space have the same cardinality).

There are many well known families of complexes and matroids. Here are some examples:

- Given a set σ, the hypergraph $\bar{\sigma}$ on σ whose edges are all subsets of σ, is called a **simplex**, and it is a complex. If $|\sigma| = d$, then σ is called a **d–simplex**. For example, the power set of $\{1, 2, 3, 4\}$ forms a 4–simplex, and can be thought of as a tetrahedron together with its faces, edges and vertices (more about this later).

- Given a graph G, the hypergraph $\mathrm{Gra}(G)$ on the edges of G whose edges correspond to subsets of edges of G that are acyclic (i.e. don't contain a cycle of G) is called the **graphic matroid** of G; for a graph G of order n, the dimension of $\mathrm{Gra}(G)$ is $n - 1$. For example, if $G = K_4$, with edges labeled as in Figure 6.1, then the graphic matroid of G has vertex set $\{a, b, c, d, e, f\}$ and bases $\{a, b, c\}, \{a, b, d\}, \{a, c, d\}, \{b, c, d\}, \{a, c, e\}$,

$\{a, d, f\}, \{b, c, f\}, \{b, d, e\}, \{a, b, e\}, \{a, b, f\}, \{c, d, e\}, \{c, d, f\}, \{a, e, f\},$
$\{b, e, f\}, \{c, e, f\}, \{d, e, f\}$ (these are precisely the spanning trees of G).

- Given a graph G, the hypergraph $\text{Cog}(G)$ on the edges of a graph G whose edges correspond to subsets X of edges of G such that $G - X$ has the same number of components as G is called the **cographic matroid** of G. If G has order n and size m, $\text{Cog}(G)$ has dimension $m - n + 1$. For example, if $G = K_4$, with edges labeled as in Figure 6.1, then the cographic matroid of G has vertex set $\{a, b, c, d, e, f\}$ and bases $\{d, e, f\},$ $\{c, e, f\}, \{b, e, f\}, \{a, e, f\}, \{b, d, f\}, \{b, c, e\}, \{a, d, e\}, \{a, c, f\}, \{c, d, f\},$ $\{c, d, e\}, \{a, b, f\}, \{a, b, e\}, \{b, c, d\}, \{a, c, d\}, \{a, b, d\}, \{a, b, c\}$ (these are precisely the complements of spanning trees of G).

- Given any subset U of a vector space W (over *any* field), the hypergraph $\text{Lin}(U)$ on U whose edges correspond to linearly independent subsets of U is a matroid.

- Given a graph G, the hypergraph $\text{Ind}(G)$ on the vertices of G whose edges correspond to subsets X of vertices that form an independent set of G is a complex (not necessarily a matroid), called the **independence complex** of G. For example, the independent set complex of C_6, as shown in Figure 6.1 has bases $\{1, 3, 5\}, \{1, 4\}, \{2, 4, 6\}, \{2, 5\}, \{3, 6\}$.

- Given a graph G, the hypergraph $\text{Cliq}(G)$ on the vertices of G whose edges correspond to subsets X of vertices that form a clique of G is a complex (not necessarily a matroid), called the **clique complex** of G. The clique complex of the graph F shown in Figure 6.1 has bases $\{1, 2\}, \{1, 5, 7\}, \{2, 3\}, \{3, 5, 6\}$.

- Given a graph G, the hypergraph $\text{Neigh}(G)$ on the vertices of G whose edges correspond to subsets X of vertices that have a common neighbour, is a complex (not necessarily a matroid), called the **neighbourhood complex** of G. The neighbourhood complex of the graph F shown in Figure 6.1 has bases $\{2, 5, 7\}, \{2, 4, 5, 6\}, \{3, 5\}, \{1, 3, 4, 6, 7\}, \{1, 5\}$.

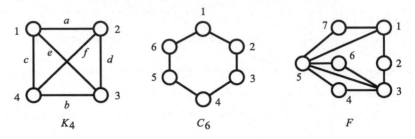

FIGURE 6.1: Graphs for complexes

What we see is that simplicial complexes are a natural way to study properties of subsets of a discrete structure, especially those that are (or can be manipulated to be) closed under containment. For the rest of this chapter, we shall assume that all combinatorial structures (including complexes) are finite (i.e. have a finite vertex set).

If complexes Δ_1 and Δ_2 are on disjoint sets X_1 and X_2, then the complex

$$\Delta_1 + \Delta_2 = \{Y_1 \cup Y_2 : Y_1 \in \Delta_1, Y_2 \in \Delta_2\}$$

is the **direct sum** of complexes Δ_1 and Δ_2. For any complex Δ and any vertex v of $\Delta = (X, E)$, we can form two subcomplexes on X. The **deletion complex** $\mathrm{del}_\Delta(v)$ has as its faces all faces τ of Δ such that $v \notin \tau$. The **link complex** $\mathrm{link}_\Delta(v)$ has as its faces all faces τ of Δ such that $v \notin \tau$ but $\{v\} \cup \tau$ is a face of Δ.

There are a few subclasses of purely dimensional complexes that are of importance to us (they have also been of interest to algebraic topologists and polytope theorists, especially the class of shellable complexes):

- A purely d–dimensional complex Δ is **vertex decomposable** if either Δ is a d–simplex, or for some vertex v of Δ, $\mathrm{link}_\Delta(v)$ is purely $(d - 1)$–dimensional and vertex decomposable, and $\mathrm{del}_\Delta(v)$ is purely d–dimensional and vertex decomposable.

- A purely d–dimensional complex Δ is **shellable** if its facets can be ordered as $\sigma_1, \ldots, \sigma_t$ with the property that for $i = 2, \ldots, t, \overline{\sigma_i} \cap (\bigcup_{j=1}^{i-1} \overline{\sigma_j})$

is purely $(d-1)$–dimensional. One can think of a shelling as a way to pull apart the complex's maximal faces along faces of one dimension lower (alternatively, by reversing the order of the shelling, one can think of a shelling as a way to build up the complex by glueing on simplices along their boundary of dimension one lower).

- A purely d–dimensional complex Δ is **partitionable** if the faces of Δ can be partitioned into intervals $[\sigma_1, \tau_1], \ldots, [\sigma_l, \tau_l]$ where each τ_i is a facet; in such a case we say that $[\sigma_1, \tau_1], \ldots, [\sigma_l, \tau_l]$ is an **interval partition** of Δ.

Example: For the cographic matroid of K_4 , the 16 bases were found to be $\{d, e, f\}$, $\{c, e, f\}$, $\{b, e, f\}$, $\{a, e, f\}$, $\{b, d, f\}$, $\{b, c, e\}$, $\{a, d, e\}$, $\{a, c, f\}$, $\{c, d, f\}$, $\{c, d, e\}$, $\{a, b, f\}$, $\{a, b, e\}$, $\{b, c, d\}$, $\{a, c, d\}$, $\{a, b, d\}$, $\{a, b, c\}$. Abbreviating a set multiplicatively, a shelling of this matroid is cde, cdf, cef, edf, bcf, bef, bcd, bef, ace, acd, aef, adf, abc, abd, abe, abf. An interval partition of the matroid is $[\emptyset, cde]$, $[f, cdf]$, $[ef, cef]$, $[edf, edf]$, $[b, bcf]$, $[be, bef]$, $[bd, bde]$, $[bef, bef]$, $[a, ace]$, $[ad, acd]$, $[ae, aef]$, $[adf, adf]$, $[ab, abc]$, $[ad, abd]$, $[abe, abe]$, $[abf, abf]$. $\triangle$

We show that vertex decomposable $\Rightarrow$ shellable $\Rightarrow$ partitionable.

Proposition 6.1 *Any vertex decomposable complex Δ is shellable.*

Proof: We proceed by induction on the construction length of Δ. If $\Delta = (X, E)$ is a simplex, then $\Delta = \overline{X}$, so X is an interval partition of Δ. Suppose now that v is a vertex of a d–dimensional vertex decomposable complex Δ such that $\operatorname{link}_\Delta(v)$ is purely $d-1$–dimensional and vertex decomposable, and $\operatorname{del}_\Delta(v)$ is purely d–dimensional and vertex decomposable. By induction, there is a shelling $\sigma_1^1, \ldots, \sigma_l^1$ of the d–dimensional complex $\operatorname{del}_\Delta(v)$ and a shelling $\sigma_1^2, \ldots, \sigma_k^2$ of the $(d-1)$–dimensional complex $\operatorname{link}_\Delta(v)$. Set $\sigma_i^3 = \sigma_i^2 \cup \{v\}$. We claim that $\sigma_1^1, \ldots, \sigma_l^1, \sigma_1^3, \ldots, \sigma_k^3$ is a shelling of Δ. For it is easy to see that they are all facets of Δ and

$$\overline{\sigma_j^3} \cap \left(\left(\bigcup_{i \le l} \overline{\sigma_i^1} \right) \cup \left(\bigcup_{r < j} \overline{\sigma_r^3} \right) \right)$$

$$= \overline{\sigma_j^2} \cup \left(\left(\overline{\sigma_j^3} \cup \overline{\sigma_1^3} \right) \cup \left(\overline{\sigma_j^3} \cup \overline{\sigma_2^3} \right) \cdots \cup \left(\overline{\sigma_j^3} \cup \overline{\sigma_{j-1}^3} \right) \right),$$

that is, $\overline{\sigma_j^2}$ together with inserted into all of the sets in $\left(\overline{\sigma_j^2} \cup \overline{\sigma_1^2} \right) \cup \left(\overline{\sigma_j^2} \cup \overline{\sigma_2^2} \right) \cdots \cup \left(\overline{\sigma_j^2} \cup \overline{\sigma_{j-1}^2} \right)$. It follows that the maximal faces are all of dimension $d - 1$. ∎

Proposition 6.2 *Any shellable complex Δ is partitionable.*

Proof: For any shelling $\sigma_1, \ldots, \sigma_t$, we define $\tau_1 = \emptyset$ and for $i \ge 2$,

$$\tau_i = \{ v \in \sigma_i : \text{ for some } j < i, \ \sigma_i - \{v\} \subseteq \sigma_j.$$

Note that $\tau_i \in \overline{\sigma_i} - \bigcup_{j < i} \overline{\sigma_j}$ as $\tau_i \subseteq \sigma_i$, and if $\tau_i \subseteq \sigma_j$ for some $j < i$, then we can extend τ_i to a maximal face τ' of $\overline{\sigma_i} \cap \bigcup_{j < i} \overline{\sigma_j}$, which is necessarily of the form $\sigma_i - \{v\}$ for some $v \in \sigma_i$, but this implies that $v \in \tau_i$, a contradiction. Moreover, if S is a face in $\overline{\sigma_i} - \bigcup_{j < i} \overline{\sigma_j}$, then for every $v \in \tau_i$, $v \in S$, as otherwise, $S \subseteq \sigma_i - \{v\} \in \overline{\sigma_i} \cap \bigcup_{j < i} \overline{\sigma_j}$, a contradiction. Thus $\tau_i \subseteq S$, so it follows that $\overline{\sigma_i} - \bigcup_{j < i} \overline{\sigma_j}$ is the interval $[\tau_i, \sigma_i]$, and that the intervals are disjoint, as $\overline{\sigma_i} - \bigcup_{j < i} \overline{\sigma_j}$ contains precisely the "new" faces of the complex $\bigcup_{j \le i} \overline{\sigma_j}$ that are not in the previous subcomplex $\bigcup_{j < i} \overline{\sigma_j}$. As every face of Δ is in $\bigcup_{j \le t} \overline{\sigma_j}$, we see that $[\tau_1, \sigma_1], \ldots, [\tau_t, \sigma_t]$ is an interval partition of Δ, so Δ is partitionable. ∎

Given a d–dimensional complex Δ, the **f–vector** of Δ is $(f_0, \ldots, f_d)$, where f_i is the number of faces of cardinality i. The **h–vector** of Δ is defined as $(h_0, \ldots, h_d)$, where h_i is given by

$$h_i = \sum_{j=0}^{i} (-1)^{i-j} \binom{d-j}{d-i} f_j$$

(h–vectors have played an important role in polytope theory, primarily with regards to the well known Upper Bound Theorem).

Example: For the cographic matroid of K_4 , the f–vector is $(1, 6, 15, 16)$ (with the labeling of K_4 in Figure 6.1, the 16 bases were found to be $\{d, e, f\}$, $\{c, e, f\}$, $\{b, e, f\}$, $\{a, e, f\}$, $\{b, d, f\}$, $\{b, c, e\}$, $\{a, d, e\}$, $\{a, c, f\}$, $\{c, d, f\}$, $\{c, d, e\}$, $\{a, b, f\}$, $\{a, b, e\}$, $\{b, c, d\}$, $\{a, c, d\}$, $\{a, b, d\}$, $\{a, b, c\}$, and all subsets of edges of size $i \leq 2$ belong to the cographic matroid as the smallest edge cut set of K_4 contains 3 edges). The h–vector can be calculated to be $(1, 3, 6, 6)$ (and it turns out that h_i counts the number of lower sets of cardinality i in an interval partition – see the previously described interval partition for the matroid). △

The h–vector of a shellable complex is always nonnegative, and has no internal zeros (the reasons for these will soon become apparent). Let

$$f(\Delta, x) = \sum_{i=0}^{d} f_i x^i$$

and

$$h(\Delta, x) = \sum_{i=0}^{d} h_i x^i$$

(these are the generating functions for the f–vector and h–vector and are called, respectively, the **f–polynomial** and **h–polynomial** of the complex Δ). It is not hard to verify that

$$h(\Delta, x) = (1 - x)^d f \left(\Delta, \frac{x}{1-x} \right), \tag{6.1}$$

and hence

$$\sum_{i=0}^{d} h_i = f_d.$$

When $\Delta = \mathrm{Cog}(G)$, the cographic matroid of a connected graph G, one sees that

$$\mathrm{Rel}(G, p) = p^m \cdot f \left(\mathrm{Cog}(G), \frac{1-p}{p} \right) = p^{n-1} \cdot h(\mathrm{Cog}(G), 1 - p)$$

and the sum of the terms in the h–vector is simply the number of spanning trees of G.

One of the classic results on complexes relates to their f–vectors. It should be clear that the components of the f–vector of a complex $\Delta = (X, E)$ are not unstructured. For example, $f_0 = 1$ and $f_1 = |X|$. Moreover, certainly if we have a lot of faces of cardinality i in our complex, there can't be too few faces of cardinality $i - 1$, as *every* subset of cardinality $i - 1$ of any face of cardinality i is a face of Δ as well. The classic bound is due to Sperner [84]:

Theorem 6.3 (Sperner's Theorem) *If* $(f_0, \ldots, f_d)$ *is the* f–vector of complex Δ on a set X of order n then we have for all $i = 1, \ldots, d$

$$f_{i-1} \geq \frac{i}{n - i + 1} f_i.$$

Proof: Consider the sets

$$\mathcal{F}_i = \{(x, \sigma) : \sigma \text{ is a face of } \Delta \text{ of cardinality } i, \ x \in \sigma\}$$

and

$$\mathcal{F}_{i-1} = \{(y, \tau) : \tau \text{ is a face of } \Delta \text{ of cardinality } i - 1, \ y \notin \tau\}.$$

We define a map

$$\rho_i : \mathcal{F}_i \to \mathcal{F}_{i-1} : (x, \sigma) \mapsto (x, \sigma - \{x\}).$$

It should be clear that indeed is a map from $\mathcal{F}_i$ into $\mathcal{F}_{i-1}$ and that the map is 1–1. This implies that

$$|\mathcal{F}_{i-1}| \geq |\mathcal{F}_i|.$$

Now $|\mathcal{F}_{i-1}| = (n - (i - 1))f_{i-1}$ while $|\mathcal{F}_i| = if_{i-1}$. Thus we conclude that

$$f_{i-1} \geq \frac{i}{n - i + 1} f_i.$$

∎

Note that argument is quite combinatorial – we find a mapping between two sets of items, each counting what we desire, and show that the map is 1–1. This simple theorem has been applied to reliability to provide simple upper and lower bounds for reliability (c.f. [31]). We have seen that the lower part of the f–vector can be determined exactly, and for the cographic matroid, the last component of the f–vector, f_{m-n+1}, is equal to the number of spanning trees, and by the **Matrix Tree Theorem**, can be calculated (as a determinant) in polynomial time. These bounds percolate up and down the f–vector via Sperner's Theorem. More accurate lower bounds for f_{i-1} in terms of f_i are found in the *Kruskal–Katona Theorem* [56], and their use in bounding can be found, for example, in [31, Section 5.4.3].

Finally, we can extend complexes to multicomplexes by allowing repetitions of elements in each face, and there is a more canonical way to represent multisets, one more in line with standard mathematical notation. Let $x_1, \ldots, x_l$ be commuting indeterminates and let $\text{Mon}(x_1, \ldots, x_l)$ denote the set of monomials in $x_1, \ldots, x_l$. The number of monomials of degree i in $\text{Mon}(x_1, \ldots, x_l)$ is

$$\binom{l-1+i}{i} = \frac{(l-1+i)!}{(l-1)!i!}$$

as such monomials are in a 1-1 correspondence with arrangements of $l-1$ identical dividers and i identical balls in a line; the number of balls to the left of the i-th divider is the power of x_i $(i = 1, \ldots, l-1)$ and the number of balls to the right of the last divider is the power of x_l.

An **order ideal of monomials** in variables $x_1, \ldots, x_l$ is a subset M of $\text{Mon}(x_1, \ldots, x_l)$ closed under division ('|'), i.e. if m_1, $m_2 \in \text{Mon}(x_1, \ldots, x_l)$, $m_1 \in M$ and $m_2 | m_1$, then $m_2 \in M$. In an obvious way, multicomplexes and order ideals of monomials are in a 1–1 correspondence, and any order ideal of square–free monomials corresponds to a complex.

If we are given $N \subseteq \text{Mon}(x_1, \ldots, x_l)$, we can define an order ideal of monomials $-N$ by 'chopping out' from $\text{Mon}(x_1, \ldots, x_l)$ all monomials that are divisible by any monomial in N, that is,

$$-N = \{m \in \text{Mon}(x_1, \ldots, x_l) : (\forall m' \in N)(m' \nmid m)\}.$$

We call N a set of **choppers** on the variables $x_1, \ldots, x_l$ for the order ideal of monomials $-N$.

Example: We can take as an order ideal of monomials on the set $\{x, y, z, w\}$ the monomials $1, x, y, z, w, x^2, y^2, z^2, xy, xz, xw, yz, yw, zw, x^3, x^2y, z^3, xyz, x^2y^2, xyzw$, as these are closed under division. A set of choppers for this order ideal of monomials is

$$\{w^2, y^3, xy^2, xz^2, x^2z, x^2w, y^2z, y^2w, z^2w, x^4, x^3y\}.$$

$\triangle$

6.1   Representations of Complexes and Multicomplexes

6.1.1   Topological Realizations of Complexes

Topologists were the first to consider complexes and to define a topology on them. For a subset X of $\mathbb{R}^n$ we let $\langle X \rangle$ denote the **convex hull** of X in $\mathbb{R}^n$, that is, $\langle X \rangle = \{\sum_{i=1}^{k} \lambda_i \mathbf{x}_i : k \geq 1, \lambda_1, \ldots, \lambda_k \in [0,1]$ and $\sum \lambda_i = 1\}$. Following Lovász and Schrijver [66] for a complex $\Delta = (X, E)$ we take a 1–1 function $\phi : X \to \mathbb{R}^n$, for some n, such that

$$\langle \phi(\sigma) \rangle \cap \langle \phi(\tau) \rangle = \langle \phi(\sigma \cap \tau) \rangle$$

for all faces σ and τ of Δ. The subspace of $\mathbb{E}^n$ on the set

$$\phi(\Delta) = \bigcup_{\sigma \in E(\Delta)} \langle \phi(\sigma) \rangle$$

is a **geometric representation** of Δ; it has the subspace topology on it. A geometric representation of $\mathcal{C}$, the complex on $V = \{1, 2, 3, 4\}$ with faces $\{\emptyset, \{1\}, \{2\}, \{3\}, \{4\}, \{1,2\}, \{1,3\}, \{1,4\}, \{2,3\}, \{2,4\}, \{3,4\}, \{1,2,3\}, \{1,2,4\}\}$, is shown in Figure 6.2.

We remark that the more standard way to describe the geometric representation of a complex Δ is as follows. A set of points $\{\mathbf{v}_0, \ldots, \mathbf{v}_k\} \subseteq \mathbb{E}^n$ are said to be **affinely independent** if there are no nontrivial solutions to the system $\lambda_0 \mathbf{v}_0 + \lambda_1 \mathbf{v}_1 + \ldots + \lambda_k \mathbf{v}_k = \mathbf{0}$, $\lambda_0 + \lambda_1 + \ldots + \lambda_k = 0$.

For a set $\{\mathbf{v}_0, \ldots, \mathbf{v}_k\} \subseteq \mathbb{E}^n$ of affinely independent points, the **Euclidean k–simplex** with vertices $\mathbf{v}_0, \ldots, \mathbf{v}_k \in \mathbb{E}^n$ is the convex hull of $\mathbf{v}_0, \ldots, \mathbf{v}_k \in \mathbb{E}^n$. A **geometric complex** $\mathcal{K}$ is the union of finitely many simplices in some $\mathbb{E}^n$

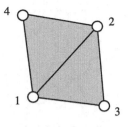

FIGURE 6.2: A complex

such that for any two of the simplices X and Y, on vertices $\{\mathbf{x}_0, \ldots, \mathbf{x}_k\}$ and $\{\mathbf{y}_0, \ldots, \mathbf{y}_l\}$ respectively, $X \cap Y$ is the convex hull of $\{\{\mathbf{x}_0, \ldots, \mathbf{x}_k\} \cap \{\mathbf{y}_0, \ldots, \mathbf{y}_l\}\}$. The abstract complex associated with this geometric complex is $\{\emptyset\} \cup \{X : \langle X \rangle \in \mathcal{K}\}$.

Clearly a complex will have infinitely many geometric representations. However, all is not lost as topologically they are all equivalent (and hence we can apply topological arguments to complexes!).

Theorem 6.4 *Any two geometric representations of a complex are homeomorphic.*

It remains to show that a complex has a geometric representation. If $\Delta = (X, E)$ is a complex with $\{v_0, v_1, \ldots, v_n\}$, and $\mathbf{e}_1, \ldots, \mathbf{e}_n$ is the standard basis in $\mathbb{R}^n$, then the map $\phi : X \to \mathbb{R}^n$ defined by $\phi(v_0) = \mathbf{0}$, $\phi(v_i) = \mathbf{e}_i$ for $i \geq 1$, defines a geometric representation of Δ. Often a complex has a representation in a Euclidean space of much lower dimension; in fact, if Δ has dimension d, it has a geometric representation in $\mathbb{R}^{2d-1}$ (c.f. [92, pg. 22]).

Scattered throughout the discrete mathematics literature there are tiny treasures of applications of geometric representations of complexes. We'll describe one briefly, a gem due to Lovász in chromatic theory.

The **Kneser graph** $G(n, k)$ is the graph on all n–subsets of $[2n + k]$, with two sets adjacent if and only if they are disjoint (the Kneser graph $G(5, 2)$ is shown in Figure 6.3). Kneser [64] conjectured that $\chi(G(n, k)) = k + 2$. Note that if we colour all of the vertices of $G(n, k)$ that contain 1 with colour 1, all of the remaining vertices that contain 2 with colour 2, $\ldots$, all of the remaining vertices that contain $k + 1$ with colour $k + 1$, and all of the vertices

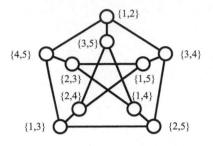

FIGURE 6.3: The Kneser graph $G(5, 2)$

left over with colour $k + 2$, then this is a good $(k + 2)$–colouring of $G(n, k)$ as at the last stage, only subsets on the set $\{k + 2, \ldots, 2n + k\}$, a set of cardinality $2n - 1$, remain, and clearly we cannot find 2 disjoint n-subsets of this set. Thus Kneser's conjecture is equivalent to showing that $G(n, k)$ is not $(k + 1)$–colourable.

Here is the idea behind Lovász' exquisite proof. First we need some topological definitions and results. The $(j + 1)$–ball $\mathcal{B}^{j+1}$ is the set of points at distance at most 1 from the origin in $\mathbb{E}^{j+1}$, and the j–sphere $\mathcal{S}^j$ is the boundary of the $(j + 1)$–ball. A topological space τ is **n–connected** if for every $j \in \{0, 1, \ldots, n\}$, every continuous map of $\mathcal{S}^j$ into τ can be extended to a continuous map of $\mathcal{B}^{j+1}$ into τ. As the property of n–connectedness is preserved under homeomorphism, we can talk about a complex being n–connected.

Lovász then considered the neighbourhood complex Neigh($G(n, k)$) of a Kneser graph. He showed essentially two things. First, he showed that the neighbourhood complex of $G(n, k)$ is $(k - 1)$–connected. Secondly, he proved the following result that related connectedness to non-colourability:

Theorem 6.5 *If Neigh(G) is $(k - 1)$–connected then G is not $(k + 1)$–colourable.*

We'll sketch the argument of this topological modeling of non k–colourability. Let ϕ be a geometric representation of the complex Neigh(G). For $\mathbf{x} \in \phi(\text{Neigh}(G))$, the vertices of the **least face of Neigh(G) containing** $\mathbf{x}$ is the smallest face σ of Neigh(G) such that $\mathbf{x} \in \langle \phi(\sigma) \rangle$.

Lovász proved the following technical result:

Lemma 6.6 ([64]) *If Neigh(G) is $(k-1)$–connected, then there is a contin-uous function $g : S^k \to \phi(Neigh(G))$ such that for all $\mathbf{x} \in \phi(Neigh(G))$, all common neighbours of the least face of Neigh(G) containing $g(\mathbf{x})$ are contained among the vertices of the least face of Neigh(G) containing $g(-\mathbf{x})$.*

Using this result, we now prove Theorem 6.5. Let G be a graph whose neighbourhood complex $\mathcal{N} = \text{Neigh}(G)$ is $(k-1)$–connected, and let g be a function as guaranteed by Lemma 6.6. Suppose – to reach a contradiction – that G had a $(k+1)$–colouring. Let S_j denote the set of points $\mathbf{x}$ of S^k such that the vertices of the least face of $\mathcal{N}$ containing $g(\mathbf{x})$ have a common neighbour coloured j. As G is connected (since $\mathcal{N}$ is 0–connected), $S_1, \ldots, S_{k+1}$ is a cover of S^k and it is easy to verify that each S_i is closed. Borsuk's Theorem (c.f. [68]) implies that one of these sets, say S_j, contains antipodal points $\mathbf{y}$ and $-\mathbf{y}$. By the definition of S_j, if U (respectively W) is the set of vertices of the least face of $\mathcal{N}$ containing $g(\mathbf{x})$ (resp. $g(-\mathbf{x})$) then there is a common neighbour u of U (resp. w of W) in G that is coloured j. By Lemma 6.6 we must have $u \in W$ and $w \in U$, so that uw is an edge of G, a contradiction as both u and w are coloured j. Thus we conclude that G is not $(k+1)$–colourable.

You might wonder why one would be interested in Kneser graphs, beyond their simple mathematical description. As a partial answer, let us define, for $k \geq 1$, the **k^{th} generalized chromatic number** of G, $\chi_k(G)$, as the least n for which there is a function f that assigns to each vertex of G a k–subset of $[n]$ such that adjacent vertices of G receive disjoint k–sets (note that $\chi_1(G) = \chi(G)$). An observation is that for any graphs G and H with $\chi(H) = k$, we have $\chi(G[H]) = \chi_k(G)$.

Garey and Johnson [46] proved the following bounds on the generalized chromatic numbers of Kneser graphs:

$$\chi_3(G(n,k)) \leq 2n+k$$

$$\chi_4(G(n,k)) \geq 2(2n+k) - 4.$$

Using these bounds and the previous observation, they proved the intriguing result that for any constant $r < 2$, if there is a polynomial–times algorithm that will colour a graph G with at most $r \cdot \chi(G)$ colours, then there is a polynomial–times algorithm that will colour a graph G with $\chi(G)$ colours.

6.1.2 Connections to Commutative Algebra

There is a way to model a complex via commutative algebra. The path is rather involved, but worth the effort to travel it, as it will show how to map a complex to a multicomplex in a way not easily describable combinatorially. To delve into it, we shall need some terminology from commutative algebra. We shall only proceed as far as necessary into the algebraic background, and refer the reader to the excellent survey [10] for a more complete discussion.

Let k be a field. A **standard graded k–algebra** is a commutative ring A containing k, and therefore a vector space over k, that is a vector space direct sum

$$A = \bigoplus_{i \geq 0} A_i$$

of subspaces $A_0, A_1 \ldots$, such that

- $A_0 = k$,

- $A_i A_j = \{a_i a_j : a_i \in A_i, a_i \in A_j\} \subseteq A_{i+j}$ for all $i, j \geq 0$, and

- there is a finite set of elements $\{z_1, \ldots, z_t\}$ of A_1 such that every element of A can be written as a polynomial in $z_1, \ldots, z_t$ with coefficients in k.

We say that the elements of A_i are **homogeneous of degree i** . The **Hilbert Series** for A,

$$\text{Hilbert}(A, x) = \sum_{i \geq 0} \text{Dim}_k(A_i) x^i,$$

is the generating series for the dimensions of the vector spaces A_i. An ideal I of a graded k–algebra A is also a subspace of A, and thus A/I is always also a vector space over k.

If I is an ideal of ring A, $z \in A$, and $\phi : A \rightarrow A/I$ is the natural homomorphism, we often write simply z for $\phi(z)$. If J' is an ideal of A/I and $J = \phi^{-1}(J')$, then clearly $(A/I)/J' \equiv A/(I + J) = A/\langle I \cup J \rangle$, and so we don't distinguish between these. We also write A (modulo $(I + J)$) for the ring $(A/I)/J'$.

We now associate a standard graded k–algebra to any complex (further details can be found, for example, in [10]). We start with a complex Δ of dimension d on a set $X = \{x_1, \ldots, x_m\}$. $k[\underline{x}] = k[x_1, \ldots, x_m]$ denotes the (commutative) polynomial ring over k in indeterminates $x_1, \ldots, x_m$. For a subset Y of X, we set

$$\sum Y = \sum_{x_i \in Y} x_i$$

and

$$\prod Y = \prod_{x_i \in Y} x_i.$$

The **Stanley–Reisner ring** of the complex Δ over k is defined as

$$k[\Delta] = k[\underline{x}]/\mathrm{Circ}(\Delta),$$

where $\mathrm{Circ}(\Delta) = \langle\{\prod Y : Y \subseteq X, \ Y \not\subseteq \Delta\}\rangle$ is the ideal generated by all subsets $Y \not\subseteq \Delta$ (clearly it suffices to take only the circuits as any set X not in Δ clearly contains a circuit of Δ). As $k[\mathbf{x}]$ is a standard graded k–algebra and $\mathrm{Circ}(\Delta)$ is generated by homogeneous polynomials, $k[\Delta]$ is also a standard graded k–algebra (see, for example, [10]). In some sense, we can think of $k[\Delta]$ as encoding Δ algebraically, as it mods out (that is, cancels out) 'non–faces' of Δ.

The standard graded k–algebra $k[\Delta]$ can be expressed as a vector space direct sum

$$k[\Delta] = \bigoplus_{i \geq 0} R_i$$

where the subspace R_i is generated by all monomials of degree i. It is easy to verify that the vector space dimension of R_i is $\sum_{j=0}^{d} \binom{i-1}{j-1} f_j$ (see Exercise 6.19). This is also the coefficient of x^i in $f(\Delta, \frac{x}{1-x})$. From

$$\mathrm{Hilbert}(k[\Delta], x) = \sum_{i \geq 0} \mathrm{Dim}_k(R_i)x^i, \tag{6.2}$$

the preceding fact and (6.1), we see that

$$h(\Delta, x) = (1 - x)^d \mathrm{Hilbert}(k[\Delta], x). \tag{6.3}$$

This connects the h–vector of a complex to the Hilbert series of the associated Stanley–Reisner ring.

A a **homogeneous system of parameters (h.s.o.p.)** is a set $\Theta = \{\theta_1, \ldots, \theta_d\}$ of homogeneous elements of degree 1 of $k[\Delta]$ if $k[\Delta]/\langle\theta_1, \ldots, \theta_d\rangle$, viewed as a vector space over k, is *finite* dimensional, that is,

$$k[\mathbf{x}]/\langle\mathrm{Circ}(\Delta) \cup \{\theta_1, \ldots, \theta_d\}\rangle = \bigoplus_{i=0}^{d} R_i'$$

where each R_i' is generated by the monomials of degree i. Stanley proved in [88] that for an *infinite* field k, a h.s.o.p. always exists for any complex Δ, (on the other hand, this can fail when the field is finite – see Exercise 6.35).

The Stanley–Reisner rings of *shellable* complexes have a certain algebraic property known as **Cohen–Macaulay** [89] (a property that is best described in terms of the vanishing of certain homological groups of the complex). It is known (see, for example, [10]) that this property implies that for any h.s.o.p. $\{\theta_1, \ldots, \theta_d\}$ of degree 1,

$$\mathrm{Hilbert}(k[\Delta]/\langle\{\theta_1, \ldots, \theta_d\}\rangle, x) \quad = \quad h_\Delta(x).$$

Stanley showed in [87] algorithmically that any quotient of a polynomial ring (and thus $k[\Delta]/\langle\{\theta_1, \ldots, \theta_d\}\rangle \cong k[\mathbf{x}]/\langle\mathrm{Circ}(\Delta) \cup \{\theta_1, \ldots, \theta_d\}\rangle$) has a vector space basis that is an order ideal of monomials, quite a surprising but useful fact. Here is a short proof, along the lines suggested by Stanley. Start with a **term ordering** of all the monomials, that is, linearly order all the monomials with 1 first and such that if $m_1 \leq m_2$ then for any monomial m, $mm_1 \leq mm_2$. (For example, a lexicographic (or reverse lexicographic) ordering on each set of monomials of the same degree, with monomials of lower degree preceding those of higher degree, is a term ordering.) Then take a large bag and successively add in monomials as long as the additions to the bag keep the bag linearly independent. It is not hard to see that the span of the monomials in the bag at the end is the span of all the monomials, and hence the entire polynomial ring. Moreover, if a monomial m is tossed into the bag at some point, and monomial m_1 divides M, say $m = m_1k$ where k is another monomial, then m_1 must already be in the bag. For if not, at the stage where

m_1 was considered for entry into the bag (and by the nature of the term ordering, this must be prior to consideration of m), it was left out as it was a linear combination of monomials already in the bag, say

$$m_1 \sum \lambda_i n_i$$

where the n_i are monomials already in the bag. However, then

$$m = m_1 k = \sum \lambda_i n_i k$$

and the monomials $n_i k$, which precede $m = m_1 k$ in the term ordering, must either be in the bag or linear combinations of items in the bag. In any event, m is a linear combination of monomials in the bag, so it can't be added, a contradiction. We conclude that the monomials in the bag, which are a basis for the polynomial ring, form an order ideal of monomials. (Another viewpoint for finding an order set of monomials that is a basis is afforded via Gröbner bases, and we will return to this later.)

From the connection to Hilbert series of Cohen–Macaulay complexes, it follows that the number of monomials of degree i in this basis is h_i, the i–component of the h–vector of Δ. This implies (via 6.13) that for a partitionable complex, the h–vector has no internal zeros, as any order ideal of monomials that has a monomial of degree i has monomials of all degrees less than i as well.

The existence of the order ideal of monomials related to the h–vector of a shellable complex Δ relies on the existence in $k[\Delta]$ of a h.s.o.p., and Stanley's arguments only allows you to conclude that such elements exist when k is infinite. Stanley in fact showed the following:

Proposition 6.7 ([88]) *Let Δ be a d–dimensional complex and*

$$\theta_1 = \sum_{i=1}^{n} a_{1,i} x_i,$$

$$\theta_2 = \sum_{i=1}^{n} a_{2,i} x_i,$$

$$\cdots$$

$$\theta_d = \sum_{i=1}^{n} a_{d,i} x_i$$

be homogeneous elements of degree 1 for $k[\Delta]$. *Then* $\{\theta_1, \ldots, \theta_d\}$ *forms a h.s.o.p. for* $k[\Delta]$ *if and only if the* $d \times n$ *matrix* $A = [a_{i,j}]$ *has the property that for every basis* σ *of* Δ, *the* $d \times d$ *submatrix of* A *formed by taking the columns corresponding to the elements of* σ *is invertible.*

If k is infinite, we have enough "wiggle room" to choose such a matrix M so that *every* $d \times d$ submatrix is nonsingular. It follows that every simplicial complex has a h.s.o.p. of degree 1 over any infinite field. Unfortunately, the argument is inherently non-constructive, and we would prefer a concrete combinatorial construction for a homogeneous system of parameters. As we will shortly show, we can do so for matroids.

We need to recall another standard definition from matroid theory (for more details, see [103]). Let Δ be a matroid on a set X of cardinality m. A **representation** over field k of Δ is a function π that maps X into some vector space V over k such that $Y \in \Delta$ if and only if $\pi(Y)$ (considered as a multiset of vectors) is linearly independent in V (that is, π preserves rank). Now if Δ is a matroid that is representable over field k, then there exists a $d \times m$ matrix M such that the submatrix consisting of d columns of M is nonsingular (i.e. of full rank) if and only if the elements of M indexing these columns form a basis of M. We deduce immediately from this and Proposition 6.7 the following.

Corollary 6.8 ([20]) *Let* Δ *be a matroid of dimension* d *over* $X = \{x_1, \ldots, x_m\}$. *Let* $\theta_1, \ldots, \theta_d$ *be any homogeneous elements of degree 1 in* $k[\Delta]$, *with* $\theta_i = \sum_{i=1}^{m} a_{i,j} x_j$. *Then* $\theta_1, \ldots, \theta_d$ *is a homogeneous system of parameters (of degree1) for* $k[\Delta]$ *if the map* $\pi : \Delta \to k^d : x_i \mapsto (a_{1,i}, \ldots, a_{d,i})^t$ *is a representation of* Δ *over* k.

Thus for a matroid Δ representable over a field k, $k[\Delta]$ has a homogeneous system of parameters of degree 1. We remark that the converse is not true, as it was observed earlier that any simplicial complex has a homogeneous system of parameters of degree 1 over any infinite field, while there are matroids that are not representable over *any* field.

One of the upshots of the algebraic connection between complexes and order ideals of monomials (i.e. multicomplexes) is that it is not apparent how to construct the order ideal of monomials under the correspondence; it is so intriguing that we have a connection drawn between two discrete structures, namely complexes and order ideals of monomials, that uses commutative algebra in a seemingly essential way.

Now we delve more deeply into ring theory and a fascinating topic in its own right, Gröbner bases (we refer the reader to [33, 48] for a survey of the theory of Gröbner bases). In short, given an ideal J of a polynomial ring $k[x_1, \ldots, x_m]$, a **Gröbner basis** is a subset of J which allows one to easily determine (among other things) whether a polynomial belongs to J.

Recall the definition of a term ordering $\preceq$ on the set of all monomials in $x_1, \ldots, x_m$, $\mathrm{Mon}(x_1, \ldots, x_m)$; $\preceq$ is a linear order such that for all monomials m, m_1, m_2, we have

- $1 \preceq m$, and

- $m_1 \preceq m_2 \rightarrow m_1 m \preceq m_2 m$.

(We have seen term orderings previously in the proof that every quotient of a polynomial ring has a basis that is an order ideal of monomials.) The largest monomial in the ordering that appears with a nonzero coefficient in a polynomial p is called the **head term** of p, and is denoted by $\mathrm{ht}(p)$; the **head coefficient**, $\mathrm{hcoeff}(p)$, is the coefficient of head term of p.

Fix a nonzero polynomial p, and set $Q \subseteq k[x_1, \ldots, x_m] - \{0\}$. We write

$$p \mapsto_Q p - \frac{sm}{\mathrm{hcoeff}(q) \cdot \mathrm{ht}(q)} q$$

if there is a polynomial $q \in Q$ such that $\mathrm{ht}(q)$ divides a monomial m in p, with the coefficient of m in p being $s \neq 0$, and say that p is **reducible modulo Q** (otherwise, p is said to be **reduced modulo Q**; the zero polynomial is also defined to be reduced as well). The idea is to use q to cancel out a monomial in p. The reflexive, transitive closure of $\mapsto_Q$ is denoted by $\mapsto^+$, and we will write $p \mapsto_Q^* r$ if $p \mapsto_Q^+ r$ and r is reduced modulo Q. A **Gröbner basis** for an ideal J is a set $GB \subseteq J$ such that for every polynomial $p \in k[x_1, \ldots, x_m]$,

$$p \in J \qquad \text{if and only if} \qquad p \mapsto_{GB}^* 0.$$

Buchberger's famous algorithm builds a Gröbner basis for an ideal J of $k[x_1, \ldots, x_m]$ (c.f. [48]). The **S–polynomial** of polynomials p and q is defined as

$$\text{Spoly}(p, q) = \text{LCM}(\text{ht}(p), \text{ht}(q)) \left(\frac{p}{\text{hcoeff}(p) \cdot \text{ht}(p)} - \frac{q}{\text{hcoeff}(q) \cdot \text{ht}(q)} \right),$$

where LCM denotes the least common multiple (of two monomials). Buchberger proved in [28] that a subset GB of ideal J is a Gröbner basis for J if and only if $\text{Spoly}(p, q) \mapsto^*_{GB} 0$ for all $p, q \in GB$.

In order to find a Gröbner basis for ideal J, Buchberger's algorithm begins with any generating set $Q = \{p_1, \ldots, p_k\}$ for the ideal. Then pairs of elements from Q are taken and their S–polynomials are reduced with respect to Q. If the result is 0, it is ignored, but otherwise it is added into Q, and the process repeats, stopping only when the S–polynomial of *any* two elements of Q reduces to 0 modulo Q (that such a process does terminate is a key feature of Buchberger's algorithm). The final set Q is a Gröbner basis for J.

It is easy to observe that the S–polynomials of two elements of an ideal J, and the reduction of any element of an ideal J with respect to subset of $J - \{0\}$, all lie in I, and hence Buchberger's algorithm extends a generating set of an ideal to a (possibly larger) generating set of J. A crucial fact (for us!) about Gröbner bases (see, for example, [48, pg. 452]) is that if GB is a Gröbner basis for ideal J of $k[x_1, \ldots, x_m]$, then (with $|$ denoting 'divides')

$$\{m \in \text{Mon}(x_1, \ldots, x_m) : (\forall m' \in GB))(\text{ht}(m') \nmid m)\}$$

is a vector space basis for $k[x_1, \ldots, x_m]/I$. Thus the head terms of the elements of a Gröbner basis are choppers (as described earlier) for the corresponding order ideal of monomials – they 'chop out' an order ideal of monomials that is a basis for $k[x_1, \ldots, x_m]/I$.

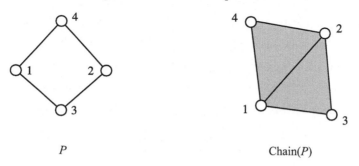

FIGURE 6.4: A partial order and its chain complex

6.2 Applications of Complexes and Multicomplexes

6.2.1 A "Complex" View of Partial Orders

A function f has a **fixed point** if $f(x) = x$ for some x in the domain of f. There has been considerable interest in determining which partial orders $P = (X, \preceq)$ have the property that every order preserving map $\gamma : V \to V$ has a fixed point; such a partial order is said to have the **fixed point property**. We shall show here how modeling partial orders with complexes can transform the problem into a topological one.

For any partial order $P = (X, \preceq)$ we can form a complex on X whose faces correspond to chains in P; this complex is called the **chain complex** Chain(P) of P.

Example: If P is a finite linear order, then Chain(P) is a simplex. If G is the partial order on the left in Figure 6.4, then its chain complex is shown on the right. $\triangle$

Two functions ρ_0 and ρ_1, each from topological spaces $\sigma = (X_1, E_1)$ to $\tau = (X_2, E_2)$, are **homotopic** if there exists a continuous map $\Phi : X_1 \times [0, 1] \to X_2$ such that $\Phi(x, 0) = \rho_0(x)$ and $\Phi(x, 1) = \rho_1(x)$ (we think of Φ as continuously deforming ρ_0 into ρ_1 within τ). We say that topological spaces $\sigma = (X_1, E_1)$ and $\tau = (X_2, E_2)$ are **homotopically equivalent** if there are continuous functions f from σ to τ and g from τ to σ such that $f \circ g$ is homotopic to the identity function on τ and $g \circ f$ is homotopic to the identity function on σ. A **contractible space** is one that is homotopic to the topological space on a single point. There are many constructions that preserve contractibility. The **Glueing Lemma** (c.f. [12, pg. 1848]) implies that if $\Delta_1 = (X_1, E_1)$, $\Delta_2 = (X_2, E_2)$ and $\Delta_1 \cap \Delta_2 = (X_1 \cap X_2, E_1 \cap E_2)$ are all contractible, then so is $\Delta_1 \cup \Delta_2 = (X_1 \cup X_2, E_1 \cup E_2)$.

A topological space $\sigma = (X, E)$ has the fixed point property if and only if for every continuous function $f : X \to X$ there is an $x \in X$ such that $f(x) = x$. Any contractible space has the fixed point property (c.f. [4, pg. 271]).

Theorem 6.9 ([4]) *Let $P = (X, \preceq)$ be a partial order. Then if a geometric representation $\phi(\mathrm{Chain}(P))$ of $\mathrm{Chain}(P)$ has the fixed point property (as a topological space) then P has the fixed point property (as a partial order).*

Proof: Let $f : X \to X$ be an order preserving map. By Exercise 6.16 we can extend f linearly to a continuous map from the topological space $\phi(\mathrm{Chain}(P))$ to itself, where f maps each $v \in X$ to $\phi(f(v))$. As in Lovász' proof of Kneser's conjecture, each point $\mathbf{x} \in \phi(\mathrm{Chain}(P))$ lies in a unique smallest simplex, say $\mathbf{x} = \sum_{i=0}^{l} \lambda_i \phi(v_i)$, where $\sum \lambda_i = 1$, all $\lambda_i > 0$ and $\{v_i : i = 1, \dots, n\}$ is a face of $\mathrm{Chain}(P)$ (that is, $\{v_i : i = 1, \dots, n\}$ is a chain of P). Without loss, $v_0 \prec v_1 \prec \dots \prec v_n$. As $\phi(\mathrm{Chain}(P))$ has the fixed point property. there is such an $\mathbf{x} \in \phi(\mathrm{Chain}(P)$ such that $f(\mathbf{x}) = \mathbf{x}$, that is,

$$\sum_{i=0}^{l} \lambda_i \phi(f(v_i)) = \sum_{i=0}^{l} \lambda_i \phi(v_i).$$

It follows that $\{v_i : i = 1, \dots, n\} = \{f(v_i) : i = 1, \dots, n\}$, so as f is order preserving and maps the chain $\{v_i : i = 1, \dots, n\}$ to itself, by Exercise 6.21 f must have a fixed point. $\blacksquare$

Thus we can sometimes look for the existence of topological fixed points of the chain complex of P when we are seeking the existence of fixed points of order preserving functions on a partial order. From the previous theorem, it follows that any partial order whose chain complex is contractible has the fixed point property. What are some examples? Partial orders with a maximum or minimum element are cones over these elements, and hence by Exercise 6.26 they are contractible. Theorem 6.9 implies that they have the fixed point property (you might try proving this directly).

Here is another example [4]. An element v of a partial order is called **irreducible** [75] if it covers exactly one element of $P - v$ or is covered by exactly one element of $P - v$. A partial order P is **dismantlable by irreducibles** if one can reduce P to a singleton by recursively removing irreducible elements. Again, it is not hard to see that the order complex of a singleton is contractible. Suppose $a_1, a_2, \ldots, a_n$ is a dismantling of P by irreducibles; we show inductively that P is contractible provided $P - a_n$ is. Assume that a_n covers exactly one element b or is covered by exactly one element b. We see that the complex Δ_2 generated by the maximal chains of P containing a_n is a cone over a_n, and hence contractible. The intersection of Δ_2 with the order complex Δ_1 of $P - a_n$ is a cone over b, and moreover $\Delta_1 \cup \Delta_2 = \text{Order}(P)$. Hence by the Glueing Lemma we conclude that $\text{Order}(P)$ is contractible, and by Theorem 6.9, it follows that P has the fixed point property.

Other contractible partial orders and other partial orders with the fixed property can be found in [12].

6.2.2 Order Ideals of Monomials and Graph Colourings

In the study of graph colourings, especially of planar graphs, the notion of a chromatic polynomial was introduced by G.D. Birkhoff back in 1912. The **chromatic polynomial** $\pi(G, x)$ of the finite graph G is the number of functions $f : V \to \{1, \ldots, x\}$ with the property that $f(u) \neq f(v)$ for all $uv \in E(G)$.

There is a **deletion–contraction** formula for calculating these functions, much like the Factor Theorem for reliability:

Theorem 6.10 *For any edge e of a graph G,*

$$\pi(G, x) = \pi(G - e, x) - \pi(G \bullet e, x).$$

Proof: Let $e = uv$ and consider the colourings of $G - e$ with x colours. These can be partitioned into (i) those that assign different colours to u and v, and (ii) those that assign the same colour to u and v. The colourings of type (i) are in 1-1 correspondence with colourings of G with x colours, and the colourings of type (ii) are in 1-1 correspondence with colourings of $G \bullet e$ with x colours. Hence

$$\pi(G - e, x) = \pi(G, x) + \pi(G \bullet e, x)$$

and the result follows. ∎

From the definition of $\pi(G, x)$ and/or via deletion-contraction and induction, we can derive the following formulas (see Exercise 6.27):

- $\pi(\overline{K_n}, x) = x^n$

- $\pi(K_n, x) = x(x - 1) \cdots (x - n + 1)$

- $\pi(T_n, x) = x(x - 1)^{n-1}$ for any tree of order n

- $\pi(C_n, x) = (x - 1)^n + (-1)^n(x - 1)$.

Example: For the graph $K_4 - e$, as shown in Figure 6.5, we find that $\pi(K_4 - e, x) = x^4 - 5x^3 + 8x^2 - 4x$. △

As $\pi(\overline{K_n}, x) = x^n$ for all n, it follows from Theorem 6.10 that $\pi(G, x)$ is indeed a polynomial, called the **chromatic polynomial** of graph G. It is a monic polynomial of degree $|V|$ in x, with integer coefficients that alternate in sign (see Exercise 6.28). Chromatic polynomials have been extremely well studied, with interest ranging from calculations to the location of their roots.

What about the combinatorial meaning, if any, of the coefficients of the chromatic polynomial in standard form, that is, written as a polynomial in terms of the standard basis $\{1, x, x^2, x^3, \ldots\}$? This will take some doing, but it will be worth the trip! We begin by finding an expansion of the chromatic polynomial in terms of spanning subgraphs.

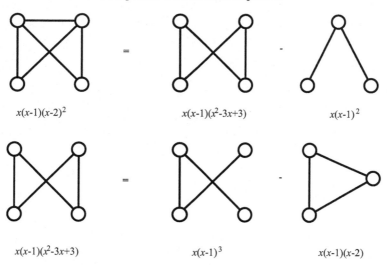

$$x(x\text{-}1)(x\text{-}2)^2 \qquad\qquad x(x\text{-}1)(x^2\text{-}3x+3) \qquad\qquad x(x\text{-}1)^2$$

$$x(x\text{-}1)(x^2\text{-}3x+3) \qquad\qquad x(x\text{-}1)^3 \qquad\qquad x(x\text{-}1)(x\text{-}2)$$

FIGURE 6.5: Calculating $\pi(K_4 - e, x)$ via deletion-contraction

Theorem 6.11 *Let G be a graph on n vertices and m edges, and let $SpSub(G)$ denote the set of all spanning subgraphs of G. Then*

$$\pi(G, x) = \sum_{H \in SpSub(G)} (-1)^{|E(H)|} x^{c(H)}$$

where $c(H)$ denotes the number of components of H.

Proof: Let the edges of G be $e_1, \ldots, e_m$. For every edge $e_i = x_i y_i$ of G, let A_i denote the set of functions $\rho : V(G) \to \{1, 2, \ldots, x\}$ such that $\rho(x_i) = \rho(y_i)$, that is, the ends of e_i get the same colour under ρ. Then clearly

$$\pi(G, x) = \left| \bigcap_{i=1}^{m} \overline{A_i} \right|$$

$$= x^n - \left| \bigcup_{i=1}^{m} A_i \right|$$

so by inclusion/exclusion we get

$$\pi(G, x) = x^n - \sum_i |A_i| + \sum_{i \neq j} |A_i \cap A_j| - \ldots + (-1)^m |A_1 \cap \cdots A_m|.$$

Now for any subset S of edges, $\bigcap_{i \in S} A_i$ consists of all functions $\rho : V(G) \to$

$\{1, 2, \ldots, x\}$ that agree on all the ends of edges in S, and this is equivalent to the set of all functions that are constant on the components of the spanning subgraph of G with edge set S. There are precisely $x^{|S|}$ of these. Note that for the spanning subgraph H' with no edges, there are $x^n = (-1)^0 x^{c(H')}$ functions that are constant on the n components of H'. So running over all choices of such edge subsets of G (or equivalently, all spanning subgraphs of G) we see that

$$\pi(G, x) = \sum_{H \in \mathrm{SpSub}(G)} (-1)^{|E(H)|} x^{c(H)}$$

and we are done. ∎

We now start to draw our big connection between chromatic polynomials and a subcomplex of the graphic matroid $\mathrm{Gra}(G)$. We start with any fixed linear order $<$ of the edge set E of G. If C is a circuit of G and $e \in C$ its $<$–least edge, then $C - e$ is called a **broken circuit** of G (with respect to order $<$). The complex on E whose faces correspond to subsets of E that do not contain a broken circuit is called the **broken circuit complex** $bc(G, <)$ of G (see, for example, [26, 27]). While the complex obviously depends on the linear order $<$ on the edge set of G, we'll see that the f–vector of the complex (which is the vital aspect of the complex for our application to chromatic polynomials) doesn't. Therefore we often abuse notation and talk about *the* broken circuit complex of G, omitting to mention the linear order. The broken broken circuit complex of a graph G is a pure complex whose dimension is equal to that of the graphic matroid, i.e. the maximum number of edges in a forest (if G is a graph of order n with c components, the dimension is thus $n - c$). An important result (see [11]) is that every broken circuit complex is shellable, though not necessarily a matroid.

Example: For the graph $K_4 - e$ shown in Figure 6.6, with the edge ordering $a < b < c < d < e$, the broken circuits are $\{b, c\}$, $\{d, e\}$ and $\{c, d, e\}$. The broken circuit complex of the graph has faces $\emptyset, \{a\}, \{b\}, \{c\}, \{d\}, \{e\}$, $\{a, b\}, \{a, c\}, \{a, d\}, \{a, e\}, \{b, d\}, \{b, e\}, \{c, d\}, \{c, e\}, \{a, b, d\}, \{a, b, e\}$. △

We return now to chromatic polynomials. Let G be a graph of order n

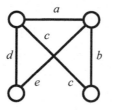

FIGURE 6.6: Graph $K_4 - e$

and size m with c components, and let $<$ any linear order of the edges. The important fact (due to Whitney [104]– see also [9, pg. 77]) is the following key result, which shows that if we expand the chromatic polynomial in the standard way, then the coefficients, ignoring signs and in reverse order, turn out to be the face numbers of the broken circuit complex (no matter which linear order $<$ is taken!).

Theorem 6.12 *Let G be a graph of order n and size m, and take a fixed linear order $<$ of the edges of G. Write*

$$\pi(G, x) = \sum_{i=0}^{n-1} (-1)^i b_i x^{n-i}. \tag{6.4}$$

Then $(b_0, \ldots, b_{n-1})$ is the f–vector of the broken circuit complex of G.

Proof: From Theorem 6.11 we have that

$$\pi(G, x) = \sum_{H \in \mathrm{SpSub}(G)} (-1)^{|E(H)|} x^{c(H)}. \tag{6.5}$$

We partition the spanning subgraphs of G, $\mathrm{SpSub}(G)$, into sets (some of which may be empty) and sum $(-1)^{|E(H)|} x^{c(H)}$ over each of the sets, and we'll see that in all but one of the cells of the partition, the sum is 0. (see Figure 6.7).

Let the circuits of G be $C_1', \ldots, C_k'$, and let f_i be the smallest edge of C_i'. Set $B_i = C_i' - \{f_i\}$, the ith broken circuit. We can assume that the broken circuits are listed so that the least edge of B_i is greater than or equal to the least edge of B_{i+1} for $i = 1, \ldots, k-1$. We now partition $\mathrm{SpSub}(G)$ into sets $\mathcal{B}_1, \mathcal{B}_1, \ldots, \mathcal{B}_k, \mathcal{B}_{k+1}$, where for $1 \leq i \leq k$, $\mathcal{B}_i$ is the set of spanning subgraphs of G that contain the broken circuit B_i but none of the previous broken circuits

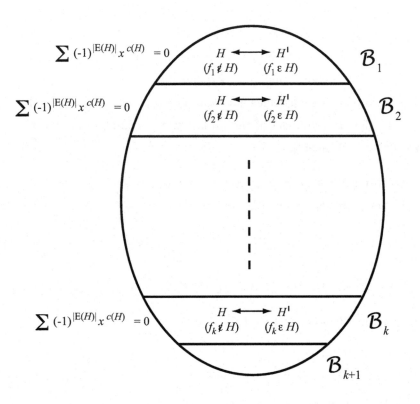

FIGURE 6.7: The partition of the spanning subgraphs of G.

$B_1, \ldots, B_{i-1}$, and $\mathcal{B}_{k+1}$ is the set of spanning subgraphs of G that contains no broken circuits (some of the sets $\mathcal{B}_1, \mathcal{B}_1, \ldots, \mathcal{B}_k, \mathcal{B}_{k+1}$ may be empty, but that will cause no difficulty).

We show now that for $1 \leq i \leq k$, the sum of $(-1)^{|E(H)|} x^{c(H)}$, over all spanning subgraphs in $\mathcal{B}_i$, is 0, by pairing off the elements of $\mathcal{B}_i$ such that the items in a pair have the same power of x but opposite signs. Consider any subgraph $H \in \mathcal{B}_i$ such that $f_i \notin H$. Let H' be the spanning subgraph of G formed by adding in edge f to H. Then $H' \in \mathcal{B}_i$ as well, as H' contains B_i, as H does, but does not contain any B_j for $j < i$, for if it did contain such a B_j, as H doesn't, B_j would necessarily have to contain f_i. But by our ordering of the broken circuits, the least edge of B_j is greater than or equal to the least edge of B_i, so f_i, being the least edge of C_i, would be smaller than the least edge of B_j, a contradiction. So the described mapping $H \to H'$ from spanning subgraphs in $\mathcal{B}_i$ not containing f_i to those that do is a mapping from $\mathcal{B}_i$ to itself. One can verify it is indeed a bijection, so it matches off elements of $\mathcal{B}_i$. Finally, under the mapping, the number of components stays the same (as adding in the edge f_i to the broken circuit B_i does not introduce a new component) but increases the number of edges by 1. Thus in the matching $(-1)^{|E(H)|} x^{c(H)}$ flips sign, cancelling in pairs. It follows that

$$\sum_{H \in \mathcal{B}_i} (-1)^{|E(H)|} x^{c(H)} = 0$$

for $i = 1, \ldots, k$.

Thus from (6.5) and our partition of $\mathrm{SpSub}(G)$ into sets $B_1, \ldots, B_{i-1}$, and $\mathcal{B}_{k+1}$, we see that

$$\pi(G, x) \quad = \quad \sum_{H \in \mathcal{B}_{k+1}} (-1)^{|E(H)|} x^{c(H)}. \tag{6.6}$$

But the edge sets of graphs in $\mathcal{B}_{k+1}$ are, by the definition of $\mathcal{B}_{k+1}$, the faces of the broken circuit complex. For each S in the broken circuit complex, the spanning subgraph H_S with edge set S is necessarily a forest (as it has no circuits). For such a set S, if $|S| = i$, then it is not hard to see that the forest

H_S has $n - i$ components. It follows from (6.5) that

$$\pi(G, x) = \sum_{i=0}^{n-1} (-1)^i b_i x^{n-i} \tag{6.7}$$

where b_i is the number of faces of cardinality i in the broken circuit complex of G, and we are done. ∎

Using (6.4) and (6.1), we can write the chromatic polynomial of a graph G in terms of the f–polynomial of the broken circuit complex and in terms of the h–polynomial of the same complex:

$$\pi(G, x) = x^n f_{bc(G,<)}(-x^{-1}) \tag{6.8}$$

$$= (-1)^{n-c} x^c (1-x)^{n-c} h_{bc(G,<)}((1-x)^{-1}). \tag{6.9}$$

This gives the underlying combinatorial (and *complex!* motivation to what is called the **forest basis expansion** of the chromatic polynomial, in terms of chromatic polynomials of forests (which are of the form $x^i(1-x)^j$).

We have just seen that the h–vector of the broken circuit complex arises as the coefficients in the forest tree expansion of the chromatic polynomial, so it is worthwhile to further investigate such h–vectors. As the broken circuit complex is shellable, by previous results we can associate an order ideal of monomials such that the sequence of number of monomials of each degree is equal to the h–vector of the complex. However, to do so, we need an explicit h.s.o.p. for the broken circuit complex of G. We observe that if Δ is the direct sum of complexes $\Delta_1, \ldots, \Delta_c$, and $\{\theta_1^i, \ldots, \theta_{d_i}^i\}$ is a h.s.o.p. for $k[\Delta_i]$ $(i = 1, \ldots, c)$, then from Proposition 6.7 we see that $\{\theta_j^i : 1 \le i \le r, 1 \le j \le d_i\}$ is a h.s.o.p. for $k[\Delta]$, since the associated matrix A of Δ described in Proposition 6.7 is block diagonal with the corresponding matrices for the Δ_i's. Thus one can form a h.s.o.p. for $k[\mathrm{Gra}(G)]$ or $k[bc(G, <)]$ by taking the unions of h.s.o.p.'s for each of G's components, so we can assume that G is a connected graph.

Let's start by fixing an orientation of the edges of G. Every minimal edge cutset (**cut**) of G is the set of edges between two sets of vertices U and W

that partition $V(G)$., We orient any such cut in one of its two directions (all edges directed from U to W, or vice versa). The **cut matrix** $D = D(G)$ is the matrix whose columns are indexed by the edges $E = \{e_1, \ldots, e_m\}$ of G, whose rows are indexed by the cuts $D_1, \ldots, D_l$ of G, and whose (i, j)–th entry is

- 1 if $e_j \in D_i$ and the orientation of e_j in G agrees with its orientation in D_i

- -1 if $e_j \in D_i$ and the orientation of e_j in G disagrees with its orientation in D_i, and

- 0 if $e_j \notin D_i$.

For a spanning tree T of G and edge $e \in T$, the **fundamental cut** $\mathcal{T}_e$ is generated by deleting e from T and taking all edges of G that join the two components of $T - e$.

Theorem 6.13 ([17]) *Let G be a connected graph of order n with edge set E. Fix a spanning tree T of G, and for each edge $e \in T$, set*

$$\theta_e = \sum_{f \in \mathcal{T}_e} d_f^e f$$

*where for $f \in \mathcal{T}_e$, d_f^e is the entry in the cut matrix D corresponding to edge f and cut $\mathcal{T}_e$. Then $\{\theta_e : e \in T\}$ is a homogeneous system of parameters for $k[Gra(G)]$ and $k[bc(G, <)]$ (for any linear order $<$ on the edges of G and **any** field k).*

Proof: We use a well known result (see Theorem 6.10 of [95], extended to any field k) that states that for any submatrix B of the cut matrix D with $n - 1$ rows and rank $n - 1$, a square $(n - 1) \times (n - 1)$ submatrix B' of B is invertible if and only if the columns of B' correspond to the edges of some spanning tree of G. This is equivalent to the columns of B being a representation of the graphic matroid $Gra(G)$ of G over k.

We define D_T to be the $(n-1) \times m$ submatrix of D whose rows correspond to the fundamental cuts of T. The matrix D_T has rank $n - 1$, since the $(n - 1) \times (n - 1)$ submatrix on the columns in T is a diagonal matrix with all diagonal entries either 1 or -1. Furthermore, F is a basis of $\mathrm{Gra}(G)$ if and only if it is a spanning tree of G, so D_T is a representation of $\mathrm{Gra}(G)$, and hence $\{\theta_e : e \in T\}$ forms a h.s.o.p. for $\mathrm{Gra}(G)$.

The shellable complexes $bc(G, <)$ and $\mathrm{Gra}(G)$ have the same dimension (namely $n - 1$). It follows that any h.s.o.p. Θ for $k[\mathrm{Gra}(G)]$ is a h.s.o.p. for $k[bc(G, <)]$ (the bases of $bc(G, <)$ are bases of $\mathrm{Gra}(G)$, and hence Stanley's method (i.e. Proposition 6.7) of determining whether a set of d elements are a h.s.o.p. shows that Θ is a h.s.o.p. for $k[bc(G, <)]$ as well). ■

To illustrate, let's consider a rather straightforward example, the cycle C_n, with edges $e_1, \ldots, e_n$. We order the edges as $e_1 < e_2 \ldots < e_n$. Clearly, the only broken circuit is $\{e_2, \ldots, e_n\}$, and taking the spanning tree $T = \{e_2, \ldots, e_n\}$, the fundamental cuts are $\{e_1, e_i\}$, $i = 2, \ldots, n$. Working over the field $\mathbb{Z}_2$ (which makes life easy, as $-1 = 1!$), we are interested in the quotient

$$\mathbb{Z}_2([e_1, e_2, \ldots, e_n])/\langle e_2 e_3 \cdots e_n, e_1 + e_2, e_1 + e_3, \ldots, e_1 + e_n\rangle$$
$$\cong \mathbb{Z}_2([e_1])/\langle e_1^{n-1}\rangle.$$

Obviously $\{1, e_1, e_1^2, \ldots, e_1^{n-1}\}$ is a basis for the latter, and this forms an order ideal of monomials. Of course, this implies that the h–vector of the broken circuit complex of C_n is $(1, 1, \ldots, 1)$, and the well known forest basis form is

$$\pi(C_n, x) = (-1)^{n-1} x \sum_{i=1}^{n-1} (1 - x)^i.$$

We turn now to a more interesting example, **wheels**. We form the n–wheel W_n ($n \geq 3$) from C_n by taking joining a new vertex to all vertices of C_n. Let the edges of C_n be $e_0, \ldots, e_{n-1}$, and the 'spokes' of W_n be labeled as $f_0, \ldots, f_{n-1}$ (see Figure 6.8) – throughout all calculations, addition of the subscripts is carried out mod n. To determine an order ideal of monomials associated with W_n, we need to find the broken circuits and the fundamental cuts with respect to some spanning tree. We fix the linear order $f_0 < \cdots <$

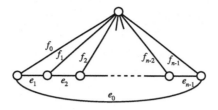

FIGURE 6.8: W_n

$f_{n-1} < e_0 < \cdots < e_{n-1}$ for the broken circuit complex. There are indeed many broken circuits in W_n, but you can verify that the minimal ones with respect to containment (and these are the ones that suffice for ideal generation) are

$$\{e_1, f_1\}, \{e_2, f_2\}, \ldots, \{e_{n-1}, f_{n-1}\}, \{e_0, f_{n-1}\},$$

$$\{f_1, e_2, \ldots, e_{n-1}, e_0\}, \{f_2, e_3, \ldots, e_{n-1}, e_0\}, \ldots, \{f_{n-2}, e_{n-1}, e_0\}$$

and

$$\{e_1, e_2, \ldots, e_{n-1}\}.$$

We make a judicious choice of a spanning tree and then determine the corresponding h.s.o.p. (once again, we choose to work over $\mathbb{Z}_2$). Let's take the spanning tree $T = \{f_i : i = 0, \ldots, n-1\}$. The corresponding h.s.o.p. (generated by the fundamental cuts) are

$$f_0 + e_0 + e_1, f_1 + e_1 + e_2, \ldots, f_{n-1} + e_{n-1} + e_0.$$

Lastly, we need to determine an order ideal of monomials that serves as a basis for the finite dimensional vector space

$$\mathbb{Z}_2[e_0, \ldots, e_{n-1}, f_0, \ldots, f_{n-1}]/I,$$

where

$$I = \langle f_0 + e_0 + e_1, f_1 + e_1 + e_2, \ldots, f_{n-1} + e_{n-1} + e_0, e_1 f_1, \ldots, e_{n-1} f_{n-1},$$

$$e_0 f_{n-1}, f_1 e_2 e_3 \cdots e_{n-1} e_0, f_2 e_3 \cdots e_{n-1} e_0, \ldots, f_{n-2} e_{n-1} e_0, e_1 e_2 \cdots e_{n-1} \rangle.$$

We start by simplifying the generators of the ideal I. Setting

$$I' = \langle f_0 + e_0 + e_1, f_1 + e_1 + e_2, \ldots, f_{n-1} + e_{n-1} + e_0, e_1^2 + e_1 e_2, e_2^2 + e_2 e_3, \ldots,$$

$$e_{n-1}^2 + e_{n-1} e_0, e_0^2 + e_{n-1} e_0, e_0^n, e_2 e_0^{n-2} + e_0^{n-1}, e_3 e_0^{n-3} + e_0^{n-2}, \ldots, e_{n-2} e_0^2 +$$

$$e_1 e_2 \cdots e_{n-1} \rangle$$

we claim that $I = I'$. To show this, we need to show that every generator of I is in I', and conversely. For example, working mod I' (i.e. in the ring $\mathbb{Z}_2[e_0, \ldots, e_{n-1}, f_0, \ldots, f_{n-1}] / I'$), we see that for any $i = 0, \ldots, n-1$, $f_i + e_i + e_{i+1} = 0$, and hence $f_i = e_i + e_{i+1}$ (don't forget that we are working over $\mathbb{Z}_2$). Hence for $i = 1, \ldots, n-1$, $e_i f_i = e_i^2 + e_i e_{i+1} = 0$, and $e_0 f_{n-1} = e_0^2 + e_{n-1} e_0 = 0$. So all of $e_1 f_1, \ldots, e_{n-1} f_{n-1}, e_0 f_{n-1}$ belong to I'. Furthermore, using $e_0^2 + e_{n-1} e_0 = 0$, we see that

$$
\begin{aligned}
f_{n-2} e_{n-1} e_0 &= (e_{n-2} + e_{n-1}) e_0^2 \\
&= e_{n-2} e_0^2 + e_0^3 \\
&= 0 \qquad\qquad (\text{mod } I')
\end{aligned}
$$

$$
\begin{aligned}
f_{n-3} e_{n-2} e_{n-1} e_0 &= (e_{n-3} + e_{n-2}) e_{n-2} e_0^2 \\
&= (e_{n-3} + e_{n-2}) e_0^3 \\
&= e_{n-3} e_0^3 + e_0^4 \\
&= 0 \qquad\qquad (\text{mod } I')
\end{aligned}
$$

$$\vdots$$

$$
\begin{aligned}
f_2 e_3 \cdots e_{n-1} e_0 &= (e_2 + e_3) e_3 \cdots e_{n-2} e_0^2 \\
&\vdots \\
&= (e_2 + e_3) e_0^{n-2} \\
&= e_2 e_0^{n-2} + e_0^{n-1} \\
&= 0 \qquad\qquad (\text{mod } I')
\end{aligned}
$$

$$\begin{aligned}
f_1 e_2 \cdots e_{n-1} e_0 &= (e_1 + e_2) e_2 \cdots e_{n-1} e_0 \\
&= e_2^2 e_3 \cdots e_{n-1} e_0 \\
&\vdots \\
&= e_2^2 e_0^{n-2} \\
&= e_0^n \\
&= 0 \qquad\qquad (\mathrm{mod}\ I').
\end{aligned}$$

Thus all the generators of I are in I', so $I \subseteq I'$. A similar argument shows the reverse inclusion, so $I = I'$.

It follows that we need to find a order ideal of monomials that form a basis for $\mathbb{Z}_2[e_0, \ldots, e_{n-1}, f_0, \ldots, f_{n-1}]/I'$. We order the monomials $\mathrm{Mon}(e_0, \ldots, e_{n-1}, f_0, \ldots, f_{n-1})$ lexicographically, where the variables are linearly ordered as

$$e_0 \preceq e_{n-1} \preceq \cdots \preceq e_1 \preceq f_0 \preceq f_1 \cdots \preceq f_{n-1}.$$

Let

$$\begin{aligned}
G' = \{ &\underline{f_0} + e_0 + e_1,\ \underline{f_1} + e_1 + e_2,\ \ldots,\ \underline{f_{n-1}} + e_{n-1} + e_0,\ \underline{e_1^2} + e_1 e_2, \\
&\underline{e_2^2} + e_2 e_3, \ldots,\ \underline{e_{n-1}^2} + e_{n-1} e_0,\ \underline{e_0^2} + e_{n-1} e_0,\ \underline{e_0^n}, e_2 e_0^{n-2} + \underline{e_0^{n-1}}, \\
&e_3 e_0^{n-3} + \underline{e_0^{n-2}}, \ldots,\ e_{n-2} e_0^2 + \underline{e_0^3},\ \underline{e_1 e_2 \cdots e_{n-1}} \},
\end{aligned}$$

so that G generates the ideal I' (the head terms of elements of G' are underlined).

By Exercise 6.20 we see that the only S-polynomials we need to calculate are:

$$\begin{aligned}
\mathrm{Spoly}(\underline{e_i^2} + e_i e_{i+1}, \underline{e_1 e_2 \cdots e_{n-1}}) &= e_{i+1} e_1 e_2 \cdots e_{n-1} \\
&\mapsto_{G'}^* 0
\end{aligned}$$

$$\begin{aligned}
\mathrm{Spoly}(\underline{e_j^2} + e_j e_{j+1}, e_0^j + \underline{e_{n-j} e_0^{n-j+1}}) &= e_j e_{j+1} e_0^{n-j} + e_j e_0^{n-j+1} \\
&\mapsto_{G'}^* e_{j+1} e_0^{n-j+1} + e_0^{n-j+2} \\
&\mapsto_{G'}^* e_0^{n-j+2} + e_0^{n-j+2} \\
&\mapsto_{G'}^* 0
\end{aligned}$$

$$\begin{aligned}
\mathrm{Spoly}(\underline{e_{n-1}^2} + e_{n-1} e_0, e_0^2 + \underline{e_{n-1} e_0}) &= e_{n-1} e_0^2 + e_{n-1} e_0^2 \\
&= 0
\end{aligned}$$

$$\text{Spoly}(\underline{e_{n-1}^2 + e_{n-1}e_0}, \underline{e_j e_0^{n-j}} + e_0^{n-j+1}) = e_j e_0^{n-j+1} + e_{n-1} e_0^{n-j+1}$$
$$\mapsto_{G'}^* \quad e_0^{n-j+2} + e_0^{n-j+2}$$
$$= \quad 0$$

$$\text{Spoly}(\underline{e_0^n}, \underline{e_j e_0^{n-j}} + e_0^{n-j+1}) = e_0^{n+1}$$
$$\mapsto_{G'}^* \quad 0$$

$$\text{Spoly}(\underline{e_j e_0^{n-j}} + e_0^{n-j+1}, \underline{e_l e_0^{n-l}} + e_0^{n-l+1}) = e_l e_0^{n-j+1} + e_j e_0^{n-j+1}$$
$$\mapsto_{G'}^* \quad e_0^{n-j+2} + e_0^{n-j+2}$$
$$= \quad 0$$

$$\text{Spoly}(\underline{e_j e_0^{n-j}} + e_0^{n-j+1}, \underline{e_1 e_2 \cdots e_{n-1}}) = e_1 e_2 \cdots e_{j-1} e_{j+1} \cdots e_{n-1} e_0^{n-j+1}$$
$$\mapsto_{G'}^* \quad e_1 e_2 \cdots e_{j-1} e_0^{n-j+2}$$
$$\mapsto_{G'}^* \quad e_1 e_0^n$$
$$\mapsto_{G'}^* \quad 0$$

for $0 \leq i \leq n-1$, $1 \leq j < l \leq n-2$. It follows that G' is in fact a Gröbner basis for ideal $I' = I$, and hence

$$\{f_0, \ldots, f_{n-1}, e_1^2, \ldots, e_{n-1}^2, \ e_0 e_{n-1}, \ e_0^n, e_0^2 e_0^{n-2}, \ e_0^3 e_0^{n-3}, \ldots, \ e_0^{n-2} e_0^2,$$
$$e_1 e_2 \cdots e_{n-1}\}$$

is a set of choppers for an order ideal of monomials that is a basis for $\mathbb{Z}_2[e_0, \ldots, e_{n-1}, \ f_0, \ldots, f_{n-1}]/I'$. This implies that the order ideal of monomials for the broken circuit complex of W_n is $\{e_0^j m : j = 0, \ldots, n-1, \ m \in \text{Mon}(e_1, \ldots, e_{n-1-j})$ is square–free and $m \neq e_1 e_2 \cdots e_{n-1}\}$.

We have, for every graph G, an associated order ideal of monomials $\text{Mon}(G)$, whose h–vector shows up in an expansion of the chromatic polynomial of G. The following theorem illustrates that there is much structure to $\text{Mon}(G)$.

Theorem 6.14 ([17]) *Let G be a connected simple graph of order n and size m, with b blocks, whose tree–basis form of its chromatic polynomial is*

$$\pi(G, x) = (-1)^{n-1} x \sum_{i=1}^{n-1} h_{n-1-i}(1-x)^i.$$

Let $M = Mon(G)$ be an order ideal of monomials associated with the broken *circuit complex of G (so that the number of monomials of degree i is precisely h_i). Then*

(i) *Mon(G) has m-n+1 monomials of degree 1 and $\binom{m-n+2}{2} - t(G)$ monomials of degree 2, where $t(G)$ is the number of triangles of G, and*

(ii) *for all i, Mon(G) has at most $\binom{m-n+i}{i}$ monomials of degree i, and if G has girth k, then Mon(G) has exactly $\binom{m-n+i}{i}$ monomials of degree i for $i < k - 1$.*

Proof: From the algebraic construction of the ideal generated by the circuits and homogeneous system of parameters, we see that for a graph G of order n and size m, a generating set of the ideal consists of $n - 1$ linear terms, each containing exactly one tree edge, and the other terms are monomials of degree at least 2. The degree of the S–polynomial of two homogeneous polynomials of degrees i and j is either 0 or has at least degree $\max(i, j)$, and the reduction procedure (when working with homogeneous polynomials) never drops the degree (unless the result is 0). Thus we see that the only choppers of degree less than two are those generated by the linear terms. Ordering the edges so that the edges of the tree are larger than the others, we see that the tree edges will be exactly the choppers of degree 1. Therefore $h_0 = 1$ and $h_1 = m - n + 1$. To calculate h_2, it is simplest just to use the definition of the h–vector.

$$
\begin{aligned}
h_2 &= \binom{n-1}{2} f_0 - \binom{n-2}{1} f_1 + \binom{n-3}{1} f_2 \\
&= \binom{n-1}{2} 1 - \binom{n-2}{1} m + \binom{n-3}{1} \left(\binom{m}{2} - t \right) \\
&= \binom{m-n+2}{2} - t
\end{aligned}
$$

where t is the number of triangles in G.

We have just seen that Mon(G) has $m - n + 1$ variables (i.e. monomials of degree 1), so it follows that Mon(G) has at most $\binom{m-n+i}{i}$ monomials of degree i. By the same reasoning as in the previous paragraph, there will be no choppers of degree less than k except for the tree edges (which are of degree 1). Thus for all $i < k$, Mon(G) has exactly $\binom{m-n+i}{i}$ monomials of degree i. ∎

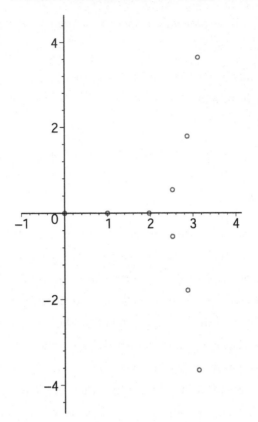

FIGURE 6.9: The chromatic roots of $K_{4,5}$

Finally, we show how much can be gained by the derived algebraic connection between complexes and order ideals of monomials. The roots of chromatic polynomials (**chromatic roots**) have attracted attention of many researchers [8, 73, 41, 16, 54, 81, 82]. On one hand, they encode the chromatic number as the first positive integer that is not a root of the polynomial, but they are interesting in and of themselves. The chromatic roots of $K_{4,5}$ are shown in Figure 6.2.2 and the chromatic roots of the theta graph $\theta_{4,4,8}$ (formed by joining two vertices by paths of lengths 4, 4 and 8) is shown in Figure 6.2.2.

In his Ph.D. thesis, Thier [94] (see also [74]) proved that for all the chromatic roots z of a graph G of order n with m edges, lie in the intersection of

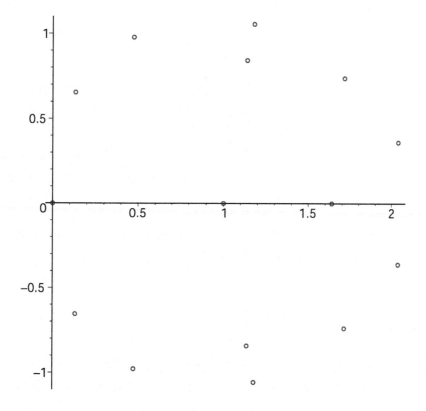

FIGURE 6.10: The chromatic roots of $\theta_{4,4,8}$

the regions

$$\{z \in \mathbb{C} : |z| \leq m - 1\} \cup \{z \in \mathbb{C} : |z - m| \leq m\},$$

$$\{z \in \mathbb{C} : |z - m + n - 2| \leq m\} \cup \{z \in \mathbb{C} : |z - 1| \leq m - 1\},$$

and

$$\{z \in \mathbb{C} : |z - 1| \leq m - 1\} \cup \{z \in \mathbb{C} : |z - m + n - 2||z - 1| \leq m(m - 1)\}.$$

These bounds are often quite weak, as, for example, the roots of

$$\begin{aligned}
\pi(C_n, x) &= (-1)^{n-1} x \sum_{i=1}^{n-1} (1 - x)^i \\
&= (-1)^{n-1} (1 - x) \left(1 - (1 - x)^{n-1}\right)
\end{aligned}$$

all lie in $|z - 1| \leq 1$. Thier's bounds were, however, the best bounds known. However, via the connection to order ideal of monomials, we can significantly improve the bound, especially for sparse graphs.

Theorem 6.15 ([17]) *Let G be a connected graph of order n and size m that is not a tree. Then the roots of $\pi(G, x)$ lie in the disc*

$$\{z \in \mathbb{C} : |z - 1| \leq m - n + 1\},$$

with equality if and only if G is unicyclic.

Proof: If G is unicyclic, then its chromatic polynomial has the form $(x - 1)^{n-i} \pi(C_i, x) = (x - 1)^{n-i} ((x - 1)^i + (-1)^i (x - 1))$, whose roots are 1 and on the circle $|z - 1| = 1$ (see Exercise 6.34). Thus we can assume that G is not unicyclic, so that $m - n + 1 \geq 2$. If G has blocks G_i of order and size n_i and m_i respectively $(i, \ldots, k)$, then $m - n + 1 = \sum_{i=1}^{k} (m_i - n_i + 1)$, and hence $m - n + 1 \geq m_i - n_i + 1$ for $i = 1, \ldots, k$. Thus we can also that G is 2–connected.

Let $(h_0, \ldots, h_{n-2})$ denote the h–vector of $bc(G, <)$. Then

$$\pi(G, x) = (-1)^n x \sum_{i=0}^{n-2} h_i (1 - x)^{n-2-i}.$$

If we set $h(z) = \sum_{i=0}^{n-2} h_i z^{n-2-i}$, it suffices to show that $h(z)$ has all its roots inside the disc centered at 0 with radius $m - n + 1$.

A well known result due to Eneström (c.f. [3, Theorem B]) says that for a polynomial $\sum a_i z^i$ with positive coefficients, all its roots lie in the disk $|z| \le \beta$, where $\beta = \max\{a_i/a_{i+1}\}$. As $h_1/h_0 = m - n + 1$, it is enough to show that

$$\frac{h_1}{h_0} \ge \frac{h_i}{h_{i-1}}, \qquad i = 2, \ldots, n - 2,$$

that is,

$$h_1 h_{i-1} \ge h_i, \qquad i = 2, \ldots, n - 2.$$

Let M be an associated order ideal of monomials for G and let M_i denote the monomials of degree i in M. We consider the set of ordered pairs

$$\mathcal{C}_{i-1} = \{(x, m_{i-1}) : x \in M_1, \ m_{i-1} \in M_{i-1}\}.$$

Now clearly $|\mathcal{C}_{i-1}| = h_1 h_{i-1}$. If we form

$$N_i = \{x m_{i-1} : (x, m_{i-1}) \in \mathcal{C}_{i-1}\},$$

then $M_i \subseteq N_i$, since if $m \in M_i$ and x is a variable that divides m, then $(x, m/x) \in \mathcal{C}_{i-1}$. Therefore

$$h_1 h_{i-1} = |\mathcal{C}_{i-1}| \ge |N_i| \ge |M_i| = h_i,$$

and we have proven the desired inequality.

Observe that if m is a monomial of degree $i \ge 2$ with at least two variables x and y, then both $(x, m/x)$ and $(y, m/y)$ belong to $\mathcal{C}_{i-1}$. If $m = x^i$, then for any variable $y \ne x$, both $(x, m/x)$ and $(y, m/y)$ belong to $\mathcal{C}_{i-1}$ as well. So if $h_1 > 1$ (i.e. G is not a cycle or an edge) then $h_1 h_{i-1} \ge 2h_i$ for $i = 2, \ldots, n-2$. That is,

$$\frac{h_1}{h_0} \ge 2\frac{h_i}{h_{i-1}} \tag{6.10}$$

for $i = 2, \ldots, n - 2$. Set

$$S = \{j = 1, \ldots, n : \frac{h_1}{h_0} > \frac{h_j}{h_{j-1}}\} \cup \{n + 1\}.$$

From Theorem 1 of [3], $h(z)$ will have a root on $|z| = m - n + 1$ if and only if the greatest common divisor of the numbers in S is greater than 1. However, from (6.10) we have $S = \{2, \ldots, n + 1\}$, and hence as G is not unicyclic $g.c.d.(S) = 1$. It follows that there is a root on the circle $|z - 1| = m - n + 1$ if and only if G is unicyclic, and we are done. ■

The bound is sharp for cycles, since all their chromatic roots lie in the disk $|z - 1| \leq 1$). It had been conjectured [16, Question 6.1] that there is a function $f : \mathbb{N} \to \mathbb{R}$ such that if G has maximum degree Δ, then all the chromatic roots of G lie in $|z| \leq f(\Delta)$. The conjecture was proved, however, in [81] for $f(\Delta) = C\Delta$, where $C \approx 8$, but it is still conjectured that perhaps even $f(\Delta) = \Delta$ may work. The previous result proves that the chromatic roots of generalized Θ–graphs (i.e those graphs with two distinguished vertices joined by three internally disjoint paths), which have maximum degree 3, all lie in $|z - 1| \leq 2$, and therefore in $|z| \leq 3$.

Exercises

Exercise 6.1 *Find the dimension of Gra(G) in terms of graph parameters of G. What are the bases of Gra(G)?*

Exercise 6.2 *Find the dimension of Cog(G) in terms of graph parameters of G. What are the bases of Cog(G)?*

Exercise 6.3 *For the graph G shown below, write out the bases for the graphic matroid, cographic matroid, independence complex, clique complex and neighbourhood complex.*

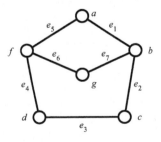

Exercise 6.4 *For the graph $K_{2,3}$, write out the the bases for the graphic matroid, cographic matroid, independence complex, clique complex and neighbourhood complex.*

Exercise 6.5 *For what graphs G is $Ind(G)$ a matroid? For what graphs G is $Cliq(G)$ a matroid?*

Exercise 6.6 *A **well–covered graph** G is one in which every maximal independent set has the same cardinality. What is this equivalent to in terms of the independence complex of G?*

Exercise 6.7 *Prove that for complexes Δ_1 and Δ_2 on disjoint sets that $\Delta_1 + \Delta_2$ is also a complex. What is the dimension of $\Delta_1 + \Delta_2$?*

Exercise 6.8 *Let Δ be a complex. Prove that indeed $del_\Delta(v)$ and $link_\Delta(v)$ are complexes.*

Exercise 6.9 *Let Δ be a d–dimensional complex. What is the dimension of $del_\Delta(v)$? What is the dimension of $link_\Delta(v)$? You will need to consider how v appears among the faces of Δ.*

Exercise 6.10 *Let Δ be a purely d–dimensional complex. Is $link_\Delta(v)$ always a pure complex? Is $del_\Delta(v)$ always a pure complex?*

Exercise 6.11 *Prove that*

$$f_\Delta(x) \quad = \quad x \cdot f_{link_\Delta(v)}(x) + f_{del_\Delta(v)}(x)$$

and for Δ, $link_\Delta(v)$ and $del_\Delta(v)$ shellable complexes of dimensions d, $d-1$ and d respectively,

$$h_\Delta(x) \quad = \quad x \cdot h_{link_\Delta(v)}(x) + h_{del_\Delta(v)}(x).$$

Exercise 6.12 *Prove that*

$$f_{\Delta_1+\Delta_2}(x) \quad = \quad f_{\Delta_1}(x) f_{\Delta_2}(x).$$

Exercise 6.13 *Suppose that Δ is a partitionable complex of dimension d with interval partition $[\sigma_1, \tau_1], \dots, [\sigma_l, \tau_l]$. Show that if $h_i = |\{j : |\sigma_l| = i\}|$, then $(h_0, h_1, \dots, h_d)$ is indeed the h–vector of Δ, and hence each h_i is a nonnegative integer. It follows that the number of lower sets of each cardinality in an interval partition of a partitionable complex is independent of the interval partition chosen.*

Exercise 6.14 *Show that any geometric representation of 2–dimensional complex Δ is a representation of the Δ as a graph in which no two edges cross except at possibly their endpoints.*

Exercise 6.15 *Show that $\{\mathbf{v}_0, \ldots, \mathbf{v}_k\} \subseteq \mathbb{E}^n$ is affinely independent if and only if $\{\mathbf{v}_1 - \mathbf{v}_0, \ldots, \mathbf{v}_k - \mathbf{v}_0\}$ is linearly independent.*

Exercise 6.16 *Show that if we have a simplicial map f between complexes $\Delta_1 = (X_1, E_1)$ and $\Delta_2 = (X_2, E_2)$ (that is, for any face $\sigma \in E_1$, $f(\sigma) \in E_2$), we can extend f (linearly) to a continuous map between geometric representations of Δ_1 and Δ_2.*

Exercise 6.17 *Prove that 0–connected is equivalent to path connected for complexes.*

Exercise 6.18 *Prove Theorem 6.5 in the case $k = 1$.*

Exercise 6.19 *Prove that the vector space dimension of R_i in (6.2) is $\sum_{j=0}^{d} \binom{i-1}{j-1} f_j$.*

Exercise 6.20 *Prove the following (see [48]).*

- *If p and q are polynomials with $q | p$, then $p \mapsto_{\{q\}} 0$.*

- *If p and q are monomials, then their S–polynomial is 0.*

- *If the head terms of p and q are relatively prime then $Spoly(p, q) \mapsto^*_{\{p,q\}} 0$.*

Thus in Buchberger's algorithm, we need only consider the S–polynomial of pairs of elements such that at least one of them is not a monomial, and their head terms are not relatively prime.

Exercise 6.21 *Prove that any finite linear order has the fixed point property.*

Exercise 6.22 *What is the dimension of the chain complex $Chain(P)$ of a partial order P (as a complex)?*

Exercise 6.23 *For any partial order* $P = (X, \preceq)$ *consider the hypergraph on X whose edges correspond to antichains in P. Show that this defines a complex, the* **antichain complex** *of P, Antichain(P). What is its dimension (as a complex)?*

Exercise 6.24 *Show that for any complex* $\Delta = (X, E)$*, one can form the partial order* $P(\Delta) = (E, \subseteq)$*. What is the height of* $P(\Delta)$*? We note that for any complex* Δ*, Order($P(\Delta)$) is the* **barycentric subdivision** *of* Δ *(c.f. [12, pg. 1844]), and any geometric representations of* Δ *and Order($P(\Delta)$) are homeomorphic.*

Exercise 6.25 *Let* $P = (X, \preceq)$ *be a partial order. Show that a function* $f : X \to X$ *is order preserving (on P) implies that the induced mapping* $f : X \to X$ *is a simplicial map on Chain(P).*

Exercise 6.26 *Prove that any geometric representation of a cone is contractible.*

Exercise 6.27 *Derive the formulas for the chromatic polynomials of empty graphs, complete graphs, trees and cycles.*

Exercise 6.28 *Prove that for a graph G of order n and size m, $\pi(G, x)$ is a monic polynomial of degree $|V|$ in x, with integer coefficients that alternate in sign.*

Exercise 6.29 *For the graph G below, find its chromatic polynomial and chromatic number.*

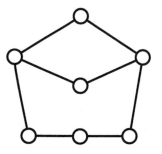

Exercise 6.30 *Describe the broken circuit complex of a tree T of order n.*

Exercise 6.31 *Describe the broken circuit complex of the cycle C_n.*

Exercise 6.32 *Prove that the real chromatic roots of a graph of order n all lie in $[0, n-1]$.*

Exercise 6.33 *Prove that the rational roots of the chromatic polynomial of graph G are precisely $0, 1, \ldots, \chi(G) - 1$.*

Exercise 6.34 *Find an expression for the roots of the chromatic polynomial of C_n, and find the closure of all such roots.*

Exercise 6.35 *Stanley [88] states that for a shellable complex Δ of dimension 2, $k[\Delta]$ has a h.s.o.p. of degree 1 if and only if Δ is $|k| + 1$–colourable (in the usual graph–theoretic sense).*
(a) Show that a complete l–partite graph (for any $l \geq 2$) is a matroid.
(b) Show that there are matroids M and finite fields k for which $k[M]$ does not have a h.s.o.p. of degree 1.

Chapter 7

Research Problems

The text has touched on a lot of interesting topics and problems, and many students may find within its pages ideas for their own research. Other students may still be on the lookout for their own problems to tackle, so I shall list a few (all are open-ended) that arise that I think are worth pursuing. All are (I hope!) accessible to a graduate student, though some may be more suitable to some students than others, depending on the mathematical and programming backgrounds of the individual.

Problem 1 *In Exercise 2.15 we discussed the use of Ehrenfeucht–Fraïssé games, as a way to examine first-order expressibility of discrete structures. The only in-depth study of the game known are those for linear orders [76] and for paths and cycles of graphs [21]. You might consider investigating the games for other discrete structures, such as partial orders or directed graphs, as well as for other interesting families of graphs.*

Problem 2 *Extend the approach taken by Straubing and Zeilberger for applying weighted directed graphs to prove results in linear algebra, and find other applications.*

Problem 3 *There are a number of results that we have seen that associate to a discrete structure a set of vectors that either linearly independent or spanning, and by calculating the dimension of the vector space, one can derive interesting new combinatorial bounds. Find some other applications of the technique, both to known combinatorial inequalities, and to some new ones!*

Problem 4 *Newton's Theorem allows one to prove unimodality of sequences by showing that the corresponding generating polynomial has only real roots.*

Find some of your own interesting applications of this result, or attempt to use it to prove partial results on established unimodality conjectures (such as those on reliability and chromatic polynomials).

Problem 5 *Finite topologies are well worth further investigation; they are finding applications in both molecular chemistry and in image analysis. Very little is known about the number of open sets of each size in a finite topology. Investigate such sequences.*

Problem 6 *Look up various properties of topological spaces, such as various forms of connectedness and compactness, and investigate their complexity on finite topologies, and whether the properties hold for most finite topologies of order n, fails for most finite topologies of order n, or neither.*

Problem 7 *A **dominating set** of a graph is a subset of the vertex set such that every vertex is either in the set or adjacent to a vertex in the set. The complements of dominating sets form a complex. Investigate the properties of such a complex. Does the f–vector have a natural and interesting interpretation?*

Problem 8 *Can stronger bounds on the roots of chromatic polynomials be derived by understanding more about the corresponding order ideal of monomials?*

Problem 9 *If we work with the field $\mathbb{Z}_2$, Buchberger's algorithm for finding a Gröbner basis seems to be inherently combinatorial; can a "useful" combinatorial algorithm be found to replace it?*

Problem 10 *Simplicial complexes have homology groups associated with them (see any textbook on algebraic topology for the definition). In a nutshell, the idea is that (if we work over a field k) one takes a free k–vector space on the faces, which naturally breaks up into the direct sum of the subspaces generated by the faces of each size. Then there are linear maps (the "boundary" maps) that carry the subspace generated by the faces of size i down into the subspace generated by faces of size $i-1$. The kernel of one map can be seen to* *contain the image of the previous map, so that one can form a sequence quotient subspaces, and it is the dimension of these subspaces that are of interest. As a project, one might investigate the combinatorial significance of homology for combinatorial properties of the complex (a few such connections have been noted in the literature in the proofs of some purely combinatorial results).*

Problem 11 *While reliability for the model of undirected graphs is well studied, there has been relatively little done for the directed case, where a digraph is considered reliable if the "up" arcs form a spanning connected subdigraph. It is not hard to see that there is an associated complex (much like the cographic matroid for the undirected case) but the complex is unfortunately not pure in general. It also appears that there is no "factor" theorem for the directed case. Investigate the properties of the associated complex.*

Selected Solutions

Chapter 2

2.3: $n!$

2.5: There are 2^n many induced structures and 2^m many spanning structures.

2.7: $r + q$ and $vq + ru^m$, respectively.

2.9: It is a preorder.

2.17: Cycles of length at least 5 have property P_1.

2.19: The mean and variance are $\binom{n}{k}p$ and $\binom{n}{k}p(1-p)$, respectively.

Chapter 3

3.1: The spectrum of the Laplacian of K_n is n ($n-1$ times) and 0 (once).

3.5: If the graph has t spanning trees and n vertices, then replacing every edge by k edges results in a multigraph with tk^{n-1} spanning trees.

3.15: $\text{Rel}(C_n, p) = (1-p) \cdot p^{n-1} + p \cdot \text{Rel}(C_{n-1}, p)$.

3.17: $-3p^6 + 14p^5 - 22p^4 + 12p^3$.

3.19: $\text{Rel}(K_{n,m}, p) = 1 - \sum \binom{n-1}{i-1}\binom{m}{j}(1-p)^{i(m-j)+(n-i)j}\text{Rel}(K_{i,j}, p)$, where the sum is taken over all $1 \leq i \leq n$, $0 \leq j \leq m$ except for $(i,j) = (n,m)$. Note that $\text{Rel}(K_{n,m}, p) = \text{Rel}(K_{m,n}, p)$. Boundary conditions are $\text{Rel}(K_{n,0}, p) = 1$ if $n = 1$, 0 otherwise, and $\text{Rel}(K_{n,1}, p) = p^{n-1}$.

Chapter 4

4.1: There are n minimal elements. If n is odd there is exactly one maximal element, otherwise there are n maximal elements. There is a maximum element iff n is odd.

4.3: No.

4.5: The minimal elements of $P_1 + P_2$ are the minimal elements of P_1, while the maximal elements of $P_1 + P_2$ are the maximal elements of P_2.

4.7: No.

4.15: (a) 1 (b) $n!$ (c) $n!m!$

4.29: $\{\emptyset, \{a\}, \{d\}, \{a,b\}, \{b,d\}, \{d,e\}, \{a,b,d\}, \{a,d,e\}, \{a,b,c,d\}, \{a,b,d,e\}, \{a,b,c$

Chapter 5

5.1: $\lceil \dfrac{n}{l-1} \rceil$.

5.27: A basis is $\{1000011, 0100101, 0010110, 0001111\}$ and the dimension is 4. A basis for the orthogonal complement is $\{0111100, 1011010, 1101001\}$.

5.29: $00000000, 1110001, 10011001, 10000111, 01010101, 00110011, 00101101,$ $01001011, 11010010, 11001100, 00011110, 10110100, 00100111, 01111000,$ $10101010, 11111111$.

Chapter 6

6.1: For a graph of order n with c components, the dimension of $\text{Gra}(G)$ is $n - c$, and the bases are the spanning forests of G.

6.3: $\text{Gra}(G)$ has bases $\{e_1, e_2, e_3, e_4, e_6\}, \{e_1, e_2, e_3, e_4, e_7\}, \{e_2, e_3, e_4, e_5, e_6\}$, $\{e_2, e_3, e_4, e_5, e_7\}, \{e_1, e_3, e_4, e_5, e_6\}, \{e_1, e_3, e_4, e_5, e_7\}, \{e_1, e_2, e_4, e_5, e_6\}$, $\{e_1, e_2, e_4, e_5, e_7\}, \{e_1, e_2, e_3, e_5, e_6\}, \{e_1, e_2, e_3, e_5, e_7\}, \{e_1, e_2, e_3, e_6, e_7\}$, $\{e_2, e_3, e_5, e_6, e_7\}, \{e_1, e_2, e_4, e_6, e_7\}, \{e_2, e_4, e_5, e_6, e_7\}, \{e_1, e_3, e_4, e_6, e_7\}$, $\{e_3, e_4, e_5, e_6, e_7\}\}$. $\text{Cog}(G)$ has bases $\{e_1, e_2\}, \{e_1, e_3\}, \{e_1, e_4\}, \{e_1, e_6\}$, $\{e_1, e_7\}, \{e_2, e_5\}, \{e_2, e_6\}, \{e_2, e_7\}, \{e_3, e_5\}, \{e_3, e_6\}, \{e_3, e_7\}, \{e_4, e_5\}, \{e_4, e_6\}$, $\{e_4, e_7\}, \{e_5, e_6\}, \{e_5, e_7\}$. $\text{Ind}(G)$ has bases $\{a, c, g\}, \{a, d, g\}, \{b, d, g\}, \{b, f\}$. $\text{Cliq}(G)$ has bases all the edges of G. $\text{Neigh}(G)$ has bases $\{b, f\}, \{a, c, g\}, \{b, d\}$, $\{c, f\}, \{a, d, g\}$.

6.5: $\text{Ind}(G)$ is a matroid iff G is the disjoint union of complete graphs. $\text{Cliq}(G)$ is a matroid iff G is a complete multipartite graph.

6.9: The dimension of $\text{del}_\Delta(v)$ is d if v is not in every basis of Δ, and is $d-1$ otherwise. The dimension of $\text{link}_\Delta(v)$ is $d-1$ if v is in every basis of Δ, and is d otherwise.

6.23: The dimension of the antichain complex of a partial order P is the width of P.

6.29: The chromatic polynomial is $x^7 - 8x^6 + 28x^5 - 55x^4 + 65x^3 - 44x^2 + 13x$ and the graph is 2–chromatic.

6.31: It consists of all subsets of a set of size n except for the whole set and one set of size $n - 1$.

Appendix A

Set Theory

The reader has undoubtedly had an introduction to sets and their notation (see, for example, [38]). The **power set** of set X is the set of subsets of X. The cardinality of a set X is denoted by $|X|$, and set X is said to have cardinality less than equal to that of Y if there is a 1–1 function $g : X \to Y$. One famous theorem is that the cardinality of the power set of a set X is always larger than the cardinality of the original set X (and hence there is no largest cardinality). A set X is **countable** if there is a 1–1 onto function $f : X \to S$ where S is finite or $S = \mathbb{N}$, the set of natural numbers; otherwise, S is **uncountable**. The real numbers $\mathbb{R}$ and the complex numbers $\mathbb{C}$ are both uncountable sets, and are much larger in size than a countable set. The removal of a countable subset S from any uncountable set X still leaves an uncountable set, in fact, one of the same cardinality as X. Even the removal of countably many subsets of X still leaves uncountably many elements left (the union of countably many countable sets is still countable). These facts are sometimes enough to show the existence of objects. For example, not every real number is the root of a polynomial with integer coefficients, as the set of such polynomials is countable, and each has a countable (in fact finite) set of roots, leading only to a countable number of roots of such polynomials. The famous continuum hypothesis states that there are no intermediate cardinalities between the size of the natural numbers and the size of the real numbers.

The **axiom of choice** states that for any set X there is a function $f :$ $\mathcal{P}(X) \to X$ such that for all nonempty subsets Y of X, $f(Y) \in Y$, that is, we can find a function that chooses an element from each nonempty subset. This axiom is equivalent to **Zorn's Lemma** which states that if $P = (V, \preceq)$ is a partial order such that every chain is bounded above, then P has a maximal element.

For sets $X_1, \ldots, X_n$, the cartesian product of $X_1, \ldots, X_n$, $X_1 \times \cdots \times X_n$, is the set of ordered n-tuples $(x_1, \ldots, x_n)$ where $x_i \in X_i$ for all $i = 1, \ldots, n$. For a positive integer m, we let $X^m = \{(x_1, \ldots, x_m) : x_i \in X\}$ denote the product of X with itself m times. There are a number of products of discrete structures whose underlying vertex set in each case is the cartesian product of the vertex sets of the structures, with various rules for constructing relations or edges from the component parts.

Given a function $f : X \to Y$ and subsets $S \subseteq X$, $T \subseteq Y$, we let $f(S) = \{f(s) : s \in S\}$ (the image of S under f) and $f^{-1}(T) = \{w : f(w) \in T\}$ (the inverse image of T under f). This notation is very handy for dealing with the action of functions on subsets. For example, in topology, a function $f : X \to Y$ is continuous if and only if the inverse image of every open set is open.

Appendix B

Matrix Theory and Linear Algebra

An elementary discussion of basic linear algebra can [101], an excellent treatment of matrices can be found be found in [57].

For an $m \times n$ matrix A with entries in some field k, the **rank** of A, $\mathrm{rank}(A)$, is the maximum number of linearly independent rows of A. A central theorem states that $\mathrm{rank}(A^T) = \mathrm{rank}(A)$, where A^T denotes the transpose of matrix A. A is said to have **full row rank** (**full column rank**) if $\mathrm{rank}(A)$ equals the number of rows (resp. columns) of A, that is, the rows (resp. columns) of A are linearly independent.

The **characteristic polynomial** of a square $n \times n$ matrix A with entries in a field $\mathbb{F}$ is $\det(\lambda I_n - A)$, where I_n denotes the $n \times n$ identity matrix; it is a polynomial of degree n in λ. The roots of the characteristic polynomial are called **eigenvalues** of A, and any nonzero vector $\mathbf{x} \in \mathbb{F}^n$ such that $A\mathbf{x} = \lambda\mathbf{x}$ is called an **eigenvector** of A (with eigenvalue λ). The **spectrum** of A is the set of eigenvalues of A, including their multiplicities as roots of the characteristic polynomial (such a multiplicity is called the algebraic multiplicity of the eigenvalue). The set $E_\lambda = \{\mathbf{v} \in \mathbb{F}^n : A\mathbf{v} = \lambda\mathbf{v}$ is called the **eigenspace** for eigenvalue λ, and its dimension as a subspace of $\mathbb{F}^n$ is called the geometric multiplicity of the eigenvalue. If a basis of $\mathbb{F}^n$ consisting of eigenvectors of A, then A is said to be **diagonalizable**, as setting P to be an $n \times n$ matrix whose columns are a basis of eigenvectors of A, we find that $P^{-1}AP$ is a diagonal matrix with the eigenvalues on the diagonal.

A real-values square matrix A is called **positive semidefinite** if all the eigenvalues of A are real and nonnegative. A theorem from matrix theory states that A is positive semidefinite iff for every vector $\mathbf{x} \in \mathbb{R}^n$, $\mathbf{x}^T A \mathbf{x} \geq 0$.

For a inner product space V over field k, two vectors $\mathbf{x}$ and $\mathbf{y}$ are said to be **orthogonal** if $\mathbf{x} \cdot \mathbf{y} = 0$ (it is not hard to see that a set of nonzero mutually orthogonal vectors are linear independent). The **norm** of a vector $\mathbf{x}$ is $||\mathbf{x}|| = \sqrt{\mathbf{x} \cdot \mathbf{x}}$. If $\mathbf{x} = (x_1, \ldots, x_n)$ and $\mathbf{y} = (y_1, \ldots, y_n)$, the usual inner product that we are interested in is $\mathbf{x} \cdot \mathbf{y} = \sum x_i y_i$. The well–known Cauchy–Schwarz inequality states that $|\mathbf{x} \cdot \mathbf{y}| \leq ||\mathbf{x}|| ||\mathbf{y}||$.

Given a vectors space V, we can form the **tensor product** $\otimes_{i=1}^n V$ as follows. We begin with the vector space V^n (the cartesian product of V with itself n times), and consider the subspace W generated by all vectors of the form

$$(x_1, \ldots, x_{i-1}, x_i + x_i', x_{i+1}, \ldots, x_n) - (x_1, \ldots, x_{i-1}, x_i, x_{i+1}, \ldots, x_n)$$
$$- \quad (x_1, \ldots, x_{i-1}, x_i', x_{i+1}, \ldots, x_n)$$

and

$$(x_1, \ldots, x_{i-1}, r x_i, x_{i+1}, \ldots, x_n) - r(x_1, \ldots, x_{i-1}, x_i, x_{i+1}, \ldots, x_n)$$

for any $i \in [n]$ and any $r \in k$. Then consider the quotient vector space V^n / w (this is a subgroup of the abelian group $(V, +)$). It is not hard to verify that the left cosets of W in V^k indeed form a vector space in the natural way: $r(\mathbf{x} + W) = r\mathbf{x} + W$ for any $x \in V^k$ and any $r \in k$. The tensor product $\otimes_{i=1}^n V$ is the vector space V^k / W. If $\{e_1, \ldots, e_d\}$ is a basis for V then $\{(e_{i_1}, \ldots, e_{i_n}) : i_j \in [d]$ for all $j\}$ is a basis for $\otimes_{i=1}^n V$. One derives the mth **exterior power** of V by taking a further quotient:

$$\wedge^k V = \otimes_{i=0}^k W / \{x_1 \otimes \ldots \otimes x_k : x_i = x_j \text{ for some } i \neq j\}.$$

Appendix C

Abstract Algebra

We assume that the reader has had introductions to linear algebra, matrices, groups, fields and ring theory. A standard reference for algebraic structures is Lang's book [58].

A group $(G, *)$ is **abelian** if $*$ is commutative, i.e. if $g * h = h * g$ for all g and h in G. If H is a subgroup of G then the **left cosets** of H in G are the sets of the form $g * H = \{g * h : h \in H\}$ (two left cosets $g_1 * H$ and $g_2 * H$ are equal iff $g_1 * g_2^{-1} \in H$). If $g_1 * H, \ldots, g_n * H$ are all of the left cosets of H in G, then $g_1, \ldots, g_n$ are called a **system of distinct representatives** for the left cosets. One can make similar definitions for right cosets, and it is not hard to show that the number of left cosets of H in G and the number of right cosets of H in G are equal (and is called the **index** of H in G). If G is abelian (more generally, if H is a normal subgroup of G) then the set of left (or right) cosets of H in G form a group, the **quotient group** G/H, where $(g_1 * H) * (g_2 * H)$ is defined to be $(g_1 * g_2) * H$ (the binary relation on the cosets is well defined due to the normality of H).

For a ring $(R, +, \cdot)$, a subset I of r is a **left ideal** (a **right ideal**) if I is a subgroup of $(R, +)$ and for all $r \in R$ and $i \in I$, we have $ri \in I$ ($ir \in I$). As $(R, +)$ is an abelian group, we can form $(R/I, +)$, and, in fact, $(R/I, +, \cdot)$ is a ring in the obvious way. Our main examples in this text of rings are the polynomial rings over a field k: we take indeterminates (i.e. variables) $x_1, \ldots, x_l$ and take all polynomials in these variables with coefficients in k (our choice for k is either a finite field $\mathbb{Z}_p$ or the infinite fields $\mathbb{Q}$ or $\mathbb{C}$). Ideals are often given by a set of **generators**, for which we take the smallest ideal containing all of the given elements.

The main algorithmic question is how to decide if a given element of the ring belongs to a given ideal. For example, if the ring is $\mathbb{Z}[x]$, then for any ideal I with generators $g_1, \ldots, g_l$, I is the ideal generated by the greatest common divisor d of $g_1, \ldots, g_l$, and hence we can easily check for membership in I. Unfortunately, in polynomial rings containing more than one variable, not every ideal is **principal**, i.e. generated by a single element) and we need a more general algorithm. We touch on a beautiful (and useful) procedure due to Buchberger in the section on complexes.

Appendix D

Probability

Our use of probability is restricted to what one might see in a first course on probability theory (see, for example, [70]). A **sample space** on a finite set Ω consists of a function $f : \Omega \to [0,1]$ such that $\sum_{\omega \in \Omega} f(\omega) = 1$. If the sample space is finite (that is, Ω is finite), then we can make Ω into a sample space by making each element of Ω equally likely, that is, by setting $\mathrm{Prob}(\omega) = 1/|\Omega|$ for each $\omega \in \Omega$. We often do just this for discrete structures. For some structures on a fixed set X it is easy to generate them with equal probability through a number of independent Bernoulli ("coin flip") trials; for example, for graphs, we can independently decide (with probability $1/2$) for each pair of vertices x and y whether to include edge (x, y) or not. For other structures, such as partial orders, it is not possible to do so, as there are dependencies among the Bernoulli trials.

An **event** A is a subset of the sample space; the complement of A, $\bar{A}$, is the set $\Omega - A$, and clearly $\mathrm{Prob}(\bar{A}) = 1 - \mathrm{Prob}(A)$. When dealing with discrete structures, events are generated by properties of interest. For example, the property of k–colourability could be an event. As events are subsets of a set, we often use set theoretic notation (such as intersection and union) to formulate compound events. One well known (and useful) result states that for events $A_1, \ldots, A_k$, we have that

$$\mathrm{Prob}\left(\bigcup_{i=1}^{k} A_i\right) \leq \sum_{i=1}^{k} \mathrm{Prob}(A_i).$$

Events $A_1, \ldots, A_k$ are **independent** if

$$\mathrm{Prob}\left(\bigcap_{i=1}^{k} A_i\right) = \prod_{i=1}^{k} \mathrm{Prob}(A_i).$$

The **conditional probability** of event A given event B is written as $\mathrm{Prob}(A|B)$ and is defined by

$$\mathrm{Prob}(A|B) = \frac{\mathrm{Prob}(A \cap B)}{\mathrm{Prob}(B)}.$$

Note that if A and B are independent then $\mathrm{Prob}(A|B) = \mathrm{Prob}(A)$. Bayes' Theorem states that for any events A and B,

$$\mathrm{Prob}(A) = \mathrm{Prob}(B)\mathrm{Prob}(A|B) + \mathrm{Prob}(\bar{B})\mathrm{Prob}(A|\bar{B}),$$

where $A|B$ denotes the conditional probability of A, given B (i.e. $\mathrm{Prob}(A|B) = \mathrm{Prob}(A \cap B)/\mathrm{Prob}(B)$).

A random variable X is a function $X : \Omega \to \mathbb{R}$. The **expectation** of the random variable X is $E(X) = \sum_{\omega \in \Omega} X(\omega)f(\omega)$, and the **variance** of X is $E((X - E(X))^2) = E(X^2) - (E(X))^2$. A simple but useful fact is that $E(X + Y) = E(X) + E(Y)$, and if the range of X is $\mathbb{N}$ and $E(X) \geq \alpha$, then $\mathrm{Prob}(X \geq \alpha) > 0$. For more basic information, see [70], and for some of the deeper connections to combinatorics, see [15].

Appendix E

Topology

Point set topology begins with the definition of a topological space: A set X together with a collection of subsets $\mathcal{O}$ of X is a **topological space** if $\emptyset \in \mathcal{O}$, $X \in \mathcal{O}$, and $\mathcal{O}$ is closed under arbitrary unions and finite intersections. The members of $\mathcal{O}$ are called the **open sets** of the topology, and the set $\mathcal{C} = \{S : X - S \in \mathcal{O}\}$ is the set of **closed sets** of the topology. A **basis** for topology $\tau = (X, \mathcal{O})$ is a collection $\mathcal{B}$ of open sets of τ such that if $x \in X$, $O \in \mathcal{O}$ and $x \in O$, then there is a $B \in \mathcal{B}$ such that $x \in B \subseteq O$.

If $\sigma = (X, \mathcal{O}_X)$ and $\tau = (Y, \mathcal{O}_Y)$ are topological spaces, then f is **continuous** if for every open set T of τ, $f^{-1}(T)$ is open in σ (equivalently, if $f^{-1}(T)$ is closed for every closed set T of τ). Two topological spaces σ and τ are **homeomorphic** if there are continuous 1–1 onto functions $f : X \to Y$ and $g : Y \to X$ such that $g(f(x)) = x$ and $f(g(y)) = y$ for all $x \in X$ and $y \in Y$ (homeomorphism is the "isomorphism" for topological spaces).

Given a topological space $\tau = (X, \mathcal{O})$ and a subset S of X, the **subspace topology** induced by S is the topology on S that has as its open sets $\{S \cap O : O \in \mathcal{O}\}$. If $\sigma = (Y, \mathcal{O}')$ is another topological space, then the **product** of τ and σ is $\tau \times \sigma = (X \times Y, \mathcal{O}'')$, where $\mathcal{O}''$ is formed by taking arbitrary unions of sets of the form $O_X \times O_Y$ where O_X and O_Y are open in τ and σ, respectively. For example, Euclidean n–space $\mathbb{E}^n$ is the product of n copies of Euclidean 1–space $\mathbb{E}^1$, the real line under its usual topology.

A topological space $\tau = (X, \mathcal{O})$ is **connected** if there are no two nonempty open sets that partition X (the connected subsets of the real line with the usual Euclidean topology are precisely the intervals). It is not hard to verify that the image of a connected space under a continuous function must be connected.

A **covering** of topological space τ on set X is a set of open sets of τ whose union is all of X; τ is **compact** if any covering of τ has a finite subcover. For example, the compact subspaces of Euclidean n–space are the closed and bounded subsets. An important theorem states that the image of a compact set under a continuous map is also compact (this is a special case of the Extreme Value Theorem from calculus).

There is so much more to study in terms of topologies. For further study, see [30].

Appendix F

Logic

Unfortunately, first–order logic is not taught as often as it used to be. Here we give a quick description of the important facts (following [29]), and leave you to pursue the topic in more detail in [29] or [6]. We assume that the reader has worked with propositional logic, and can create truth tables for the usual connectives $\neg$ (not), $\wedge$ (and), $\vee$ (or) and $\rightarrow$ (implication).

Let $\mathcal{C}$ be a set of **constants** and let X be a set of **variables**. The **terms** are the constants and variables. A **predicate** is of the form $P(x_1, \ldots, x_n)$ where P is a symbol, n is a positive integer (called the **arity** of the predicate) and $x_1, \ldots, x_n$ are terms. Let $\mathcal{P}$ be a set of **predicates** (including the special binary predicate of equality, $=$) that have their own arity (these are thought of as abstractions of relations), The **atomic formulae** are those formulas of the form $s = t$ where s and t are terms, and $P(s_1, \ldots, s_n)$ where the s_i are terms and P is an n–ary predicate. The **formulae** of the language $\mathcal{L}$ are defined recursively: all atomic formulae are formulae, if α and β are formulae then so are $\neg\alpha$, $\alpha \wedge \beta$, $\alpha \vee \beta$, $\alpha \rightarrow \beta$, and $\forall(x)\alpha$, where x is any variable in the language. The symbol $\forall$ is read as "for all", and we introduce the shorthand $\exists$ for $\neg\forall\neg$ (and is read "there exists"); these two symbols are called **quantifiers**. The **sentences** are the formulae without "free" (i.e. unquantified) variables, and these sentences form a **language**.

A **theory** Γ is a set of sentences in a language. A **proof** of a sentence ϕ from a theory Γ is a sequence of lines, where the last line is ϕ, and each line is either in Γ or follows from the previous ones by one of a short list of deductive rules for first–order logic, the details of which can be found, for example, in [29] or [6] (different authors use different rules, but they all amount to the same thing). If we can prove ϕ from Γ, we write $\Gamma \vdash \phi$. The set $\mathrm{Ded}(\Gamma)$ is the set of all sentences that can be proved (or "deduced") from Γ. Axioms for a theory Γ are a set A of sentences for which $\mathrm{Ded}(A) = \mathrm{Ded}(\Gamma)$ (the theory is **finitely axiomatizable** if it has a finite set of axioms). The theory Γ is **consistent** if you cannot prove $p \wedge \neg p$ from Γ for some sentence p (if Γ is inconsistent, then *any* sentence can be proved from Γ). As any proof uses only finitely many sentences, it follows that a theory Γ is consistent iff every finite subset of Γ is consistent. A theory Γ is **complete** if for every sentence ϕ in the language, either $\Gamma \vdash \phi$ or $\Gamma \vdash \neg\phi$.

If proofs are "abstract", then "models" are concrete. A **model** $\mathcal{M}$ of a theory Γ consists of a set D, called a **domain**, and interpretations of the constants as elements in D, and relations of the proper arity for the predicates. The sentences are interpreted in the obvious way (for example, $\forall(x)(P(x, y)$ is true in the model $\mathcal{M}$ if $R(a, b)$ is true for every a and b in D, where R is the interpretation of P in $\mathcal{M}$. Γ is **satisfiable** if it has a model. For a theory Γ and a sentence ϕ, we write $\Gamma \models \phi$ if ϕ is true in every model in which all of Γ true.

Two models $\mathcal{M}$ and $\mathcal{N}$, with domains $D_{\mathcal{M}}$ $D_{\mathcal{N}}$ respectively, of a theory Γ are **isomorphic** if there is a 1–1 onto function $f : D_{\mathcal{M}} \to D_{\mathcal{N}}$ between their domains such that for every constant c, the interpretations of c in both models are matched to one another by f, and for every n–ary predicate P in the language, if R and R' are the interpretations of P in $\mathcal{M}$ and $\mathcal{N}$ respectively, then for all $x_1, \ldots, x_m \in D_{\mathcal{M}}$, $R(x_1, \ldots, x_m)$ is true iff $R'(f(x_1), \ldots, f(x_m))$ is true. If two models of a theory are isomorphic, then exactly the same sentences are true in both models.

A major result in first-order logic is that the equivalence of the abstract and concrete. Namely, a theory Γ is consistent iff it is satisfiable. Moreover, for any sentence *phi*, $\Gamma \vdash \phi$ iff $\Gamma \models \phi$. From this connection we deduce a well-known theorem that bridges the gap from the finite to the infinite:

Theorem F.1 (The Compactness Principle) *A set of sentences in first-order logic has a model if and only if every finite subset has a model.*

Finally, the **Löwenheim–Skolem Theorem** states that a consistent countable theory has a *countable* model.

Bibliography

[1] B.A. Anderson. Families of mutually complementary topologies. *Proc. Amer. Math. Soc.*, 29:362–368, 1971.

[2] B.A. Anderson. Finite topologies and hamiltonian paths. *J. Combin. Theory Ser. B*, 14:87–93, 1973.

[3] N. Anderson, E.B. Saff, and R.S. Varga. On the Eneström–Kakeya theorem and its sharpness. *Lin. Alg. Appl.*, 28:5–16, 1973.

[4] K. Backlawski and A. Björner. Fixed points in partially ordered sets. *Adv. Math.*, 31:263–287, 1979.

[5] M.O. Ball and J.S. Provan. Bounds on the reliability polynomial for shellable independence systems. *SIAM J. Alg. Disc. Meth.*, 3:166–181, 1982.

[6] D.W. Barnes and J.M. Mack. *An algebraic introduction to mathematical logic*. Springer, New York, 1975.

[7] L. Batten. *Combinatorics of finite geometry*. Cambridge Univ. Press, New York, 1986.

[8] G. Berman and W.T. Tutte. The golden root of a chromatic polynomial. *J. Comb. Th.*, 6:301–302, 1982.

[9] N. Biggs. *Algebraic graph theory*. Cambridge Univ. Press, Cambridge, 1993.

[10] L.J. Billera. Polyhedral theory and commutative algebra. In *Mathematical programming: the state of the art (A. Bachem, M. Grötchel and B.H. Korte, ed.)*, pages 57–77. Springer, New York, 1983.

[11] A. Björner. Homology and shellability. In *Homology and Shellability (N. White, ed.)*, pages 226–283. Cambridge University Press, Cambridge, 1992.

[12] A. Björner. Topological methods. In *Handbook of combinatorics II (R.L. Graham et al, ed.)*, pages 226–283. Elsevier, Cambridge, 1995.

[13] B. Bollobás. On generalized graphs. *Acta Math. Acad. Sci. Hungar.*, 16:447–452, 1965.

[14] B. Bollobás. *Graph theory.* Springer, New York, 1979.

[15] B. Bollobás. *Random graphs.* Academic Press, London, 1985.

[16] F. Brenti, G.F. Royle, and D.G. Wagner. Location of zeros of chromatic and related polynomials of graphs. *Canad. J. Math.*, 46:55–80, 1994.

[17] J.I. Brown. Chromatic polynomials and order ideals of monomials). *Discrete Math.*, 189:43–68, 1998.

[18] J.I. Brown and C.J. Colbourn. Roots of the reliability polynomial. *SIAM J. Disc. Math.*, 5:571–585, 1992.

[19] J.I. Brown and C.J. Colbourn. On the log concavity of reliability and matroidal sequences. *Adv. Appl. Math.*, 15:114–127, 1994.

[20] J.I. Brown, C.J. Colbourn, and D.G. Wagner. Cohen–Macaulay rings in network reliability. *SIAM J. Disc. Math.*, 3:377–392, 1996.

[21] J.I. Brown and R. Hoshino. The Ehrenfeucht–Fraisse game for paths and cycles. *Ars Combin.*, 8:193–212, 2007.

[22] J.I. Brown and V. Rödl. A Ramsey type problem concerning vertex colourings. *J. Comb. Th. (B)*, 52:45–52, 1991.

[23] J.I. Brown and S. Watson. Mutually complementary partial orders. *Discrete Math.*, 113:27–39, 1993.

[24] J.I. Brown and S. Watson. Partial order complementation graphs. *Order*, 11:237–255, 1994.

[25] J.I. Brown and S. Watson. The number of complements of a topology on n points is at least 2^n (except for some special cases). *Discrete Math.*, 154:27–39, 1996.

[26] T. Brylawski. The broken circuit complex. *Trans. Amer. Math. Soc.*, 234:417–433, 1977.

[27] T. Brylawski and J. Oxley. The broken circuit complex: its structure and factorizations. *Europ. J. Combinatorics*, 2:107–121, 1981.

[28] B. Buchberger. A theoretical basis for the reduction of polynomials to canonical forms. *ACM SIGSAM Bull.*, 10:19–29, 1976.

[29] C.C. Chang and H.J. Keisler. *Model theory.* North-Holland, Amsterdam, 1977.

[30] C.O. Christenson and W.L. Voxman. *Aspects of topology.* Marcel Dekker, New York, 1977.

[31] C.J. Colbourn. *The combinatorics of network reliability.* Oxford Univ. Press, New York, 1987.

[32] L. Comtet. *Advanced combinatorics.* Reidel Pub. Co., Boston, 1974.

[33] D. Cox, J. Little, and D. O'Shea. *Ideals, varieties, and algorithms.* Springer, New York, 1992.

[34] N. de Bruijn and P. Erdös. A colour problem for infinite graphs and a problem in the theory of relations. *Indig. Math.*, 13:369–373, 1951.

[35] J.-P. Doignon, A. Ducamp, and J.-C. Falmagne. On realizable biorders and the biorder dimension of a relation. *J. Math. Psych.*, 28:73–109, 1984.

[36] P. Edrös. Graph theory and probability. *Canad. J. Math.*, 11:34–38, 1959.

[37] A. Ehrenfeucht. An application of games to the completeness problem for formalized theories. *Fund. Math.*, 49:129–141, 1961.

[38] H. Enderton. *Elements of set theory.* Academic Press, New York, 1977.

[39] M. Erné. Struktur– und anzahlformeln fur topologien auf endlichen mengen. *Manuscripta Math.*, 11:221–259, 1961.

[40] J.W. Evans, F. Harary, and M.S. Lynn. On the computer enumeration of finite topologies. *Comm. ACM*, 10:295–297, 1967.

[41] E.J. Farell. Chromatic roots – some observations and conjectures. *Disc. Math.*, 29:161–167, 1980.

[42] R.A. Fisher. An examination of the possible different solutions of a problem in incomplete blocks. *Ann. Eugenics*, 10:52–75, 1940.

[43] J. Folkman. Graphs with monochromatic complete subgraphs in every edge coloring. *SIAM J. Appl. Math.*, 18:19–24, 1970.

[44] R. Fraïssé. Sur quelques classifications des systèmes de relations. *Publications Scientifiques de l'Université d'Alger, Ser. A*, 1:35–182, 1954.

[45] S. Franklin and Y. Zalcstein. Testing homotopy equivalence is isomorphism complete. *Discrete Appl. Math.*, 13:101–104, 1986.

[46] M.R. Garey and D.S. Johnson. The complexity of near–optimal graph coloring. *J. ACM*, 23:43–49, 1976.

[47] M.R. Garey and D.S. Johnson. *Computers and intractibility.* W.H. Freeman, New York, 1979.

[48] K.O. Geddes, S.R. Czapor, and G. Labahn. *Algorithms for computer algebra.* Kluwer, Boston, 1992.

[49] A. Ghoula-Houri. Caractérisation des graphs non orientes dont on peut orienter les arêtes de manière à obtenir le graphe d'une relation d'ordre. *C.R. Academic Sci. Paris*, 254:1370–1371, 1962.

[50] M.C. Golumbic. *Algorithmic graph theory and perfect graphs.* Academic Press, New York, 1980.

[51] G. Hardy, J.E. Littlewood, and G. Pólya. *Inequalities.* Cambridge Univ. Press, Cambridge, 1952.

[52] J. Hartmanis. On the lattice of topologies. *Canad. J. Math.*, 10:547–553, 1958.

[53] S.G. Hoggar. Chromatic polynomials and logarithmic concavity. *J. Comb. Th. B*, 16:248–254, 1974.

[54] B. Jackson. A zero–free interval for chromatic polynomials of graphs. *Comb. Prob. and Comp.*, 2:325–336, 1993.

[55] D. Kleitman and B. Rothschild. The number of finite topologies. *Proc. Amer. Math. Soc.*, 25:276–282, 1970.

[56] J.B. Kruskal. The number of simplices in a complex. In *Mathematical optimization techniques (R. Bellman, ed.)*, pages 251–278. Univ. of California Press, 1963.

[57] P. Lancaster and M. Tismenetsky. *The theory of matrices.* Academic Press, New York, 1985.

[58] S. Lang. *Linear algebra (2nd ed.).* Addison Wesley, London, 1972.

[59] R.E. Larson and S. Andima. The lattice of topologies: a survey. *Rocky Mountain J. Math.*, 5:177–198, 1975.

[60] L. Lovász. On chromatic number of finite set-systems. *Acta Math. Academic Sci. Hungar.*, 19:59–67, 1968.

[61] L. Lovász. Covering and Colorings of Hypergraphs. In *Proceedings of the 4th Southeastern Conference on Combinatorics*, volume 3, pages 3–12. Utilitas Math., 1973.

[62] L. Lovász. Flats in matroids and geometric graphs. In *Combinatorial surveys (Proc. 6th British Comb. Conf.)*, pages 45–86. Academic Press, New York, 1977.

[63] L. Lovász. A homology theory for spanning tress of a graph. *Acta Math. Academic Sci. Hungar.*, 30:241–251, 1977.

[64] L. Lovász. Kneser's conjecture, chromatic number and homotopy. *J. Comb. Th. A*, 25:319–324, 1978.

[65] L. Lovász. Topological and algebraic methods in graph theory. In *Graph theory and related topics (J.A. Bondy, U.S.R. Murty, eds.)*, pages 1–14. Academic Press, New York, 1979.

[66] Lovász, L. and Schrijver, A. manuscript.

[67] F.J. MacWilliams and N.J.A. Sloane. *The theory of error-correcting codes*. North-Holland, New York, 1977.

[68] J. Matoušek. *Using the Borsuk-Ulam theorem*. Springer, Berlin, 2003.

[69] J. Nešetřil and V. Rödl. Partitions of vertices. *Comm. Math. Univ. Carolinae*, 17:85–95, 1976.

[70] M. Nosal. *Basic probability and applications*. Saunders, Philadelphia, 1977.

[71] I. Rabinovitch and I. Rival. The rank of distributive lattice. *Discrete Math.*, 25:275–279, 1979.

[72] R. Read. An introduction to chromatic polynomials. *J. Comb. Th.*, 4:52–71, 1968.

[73] R. Read and G.F. Royle. Chromatic roots of families of graphs. In *Graph theory, combinatorics and applications*, pages 1009–1029. Wiley, New York, 1991.

[74] R. Read and W.T. Tutte. Chromatic polynomials. In *Selected topics in graph theory 3 (L.W. Beineke and R.J. Wilson, ed.)*, pages 15–42. Academic Press, New York, 1988.

[75] I. Rival. A fixed point theorem for finite partially ordered sets. *J. Comb. Th. A*, 21:309–318, 1976.

[76] J. Rosenstein. *Linear orderings*. Academic Press, New York, 1982.

[77] P.S. Schnare. Multiple complementation in the lattice of topologies. *Fund. Math.*, 62:53–59, 1968.

[78] P.S. Schnare. Infinite complementation in the lattice of topologies. *Fund. Math.*, 64:249–255, 1969.

[79] J.J. Seidel. *Introduction to association schemes*. Séminaire Lotharingien de Combinatoire, Thurnau, 1991.

[80] P.D. Seymour. On the two-colouring of hypergraphs. *Quart. J. Math. Oxford*, 25:303–312, 1974.

[81] A.D. Sokal. Bounds on the complex zeros of (di)chromatic polynomials and Potts-model partition functions. *Combin. Probab. Comput.*, 10:41–77, 2001.

[82] A.D. Sokal. Chromatic roots are dense in the whole complex plane. *Combin. Probab. Comput.*, 13:221–261, 2004.

[83] A.D. Sokal and G. Royle. The Brown-Colbourn conjecture on zeros of reliability polynomials is false. *J. Comb. Th. B*, 13:345–360, 2004.

[84] E. Sperner. Ein satz über untermengen einer endlichen menge. *Math. Zeit.*, 27:544–548, 1928.

[85] R.P. Stanley. On the number of open sets of finite topologies. *J. Comb. Th.*, 10:74–79, 1971.

[86] R.P. Stanley. An extremal problem for finite topologies and distributive lattice. *J. Comb. Th. A*, 14:209–214, 1973.

[87] R.P. Stanley. Cohen-Macaulay complexes. In *Higher combinatorics (M. Aigner, ed.)*, pages 51–62. Reidel, Boston, 1977.

[88] R.P. Stanley. Balanced Cohen–Macaulay complexes. *Trans. Amer. Math. Soc.*, 249:139–157, 1979.

[89] R.P. Stanley. *Enumerative combinatorics vol. 1.* Birkhäuser, Boston, 1983.

[90] A.K. Steiner. The lattice of topologies: structure and complementation. *Trans. Amer. Math. Soc.*, 122:379–397, 1966.

[91] D. Stephen. Topology on finite sets. *Amer. Math. Monthly*, 75:739–741, 1968.

[92] J. Stillwell. *Classical topology and combinatorial group theory.* Springer, New York, 1980.

[93] H. Straubing. A combinatorial proof of the Cayley–Hamilton theorem. *Discrete Math.*, 27:273–279, 1983.

[94] V. Thier. *Graphen and Polynome (Ph.D. thesis).* TU München, München, 1983.

[95] K. Thulasiraman and M.N.S. Swamy. *Graphs: theory and algorithms.* Wiley, New York, 1992.

[96] B. Toft. Colour–critical graphs and hypergraphs. *J. Comb. Th. B*, 16:145–161, 1974.

[97] W.T. Trotter. *Combinatorics and partially ordered sets.* Johns Hopkins, Baltimore, 1992.

[98] W.T. Trotter and J. Moore. The dimension of planar posets. *J. Comb. Th. B*, 21:51–67, 1977.

[99] H. Tverberg. On the decomposition of k_n into complete bipartite graphs. *J. Graph Th.*, 6:493–494, 1982.

[100] A.C.M. van Rooij. The lattice of all topologies is complemented. *Canad. J. Math.*, 20:805–807, 1968.

[101] F. Venit, W. Bishop, and J.I. Brown. *Elementary linear algebra (First Canadian Edition).* Nelson Canada, Toronto, 2008.

[102] D.G. Wagner. Zeros of reliability polynomials and f–vectors of matroids. *Combin. Probab. Comput.*, 9:167–190, 2000.

[103] D.J.A. Welsh. *Matroid theory.* Academic Press, London, 1976.

[104] H. Whitney. A logical expansion in mathematics. *Bull. Amer. Math. Soc.*, 38:572–579, 1932.

[105] R.M. Wilson. An existence theorem for pairwise balanced designs ii: The structure of pbd-closed sets and the existence conjectures. *J. Comb. Th. A*, 13:246–273, 1971.

[106] D. Zeilberger. A combinatorial approach to matrix theory. *Discrete Math.*, 56:61–72, 1985.

Index